Advances in Experimental Philosophy of Logic and Mathematics

Advances in Experimental Philosophy

Series Editor:

James R. Beebe, Professor of Philosophy, University at Buffalo, USA

Editorial Board:

Joshua Knobe, Yale University, USA
Edouard Machery, University of Pittsburgh, USA
Thomas Nadelhoffer, College of Charleston, UK
Eddy Nahmias, Georgia State University, USA
Jennifer Nagel, University of Toronto, Canada
Joshua Alexander, Siena College, USA

Empirical and experimental philosophy is generating tremendous excitement, producing unexpected results that are challenging traditional philosophical methods. *Advances in Experimental Philosophy* responds to this trend, bringing together some of the most exciting voices in the field to understand the approach and measure its impact in contemporary philosophy. The result is a series that captures past and present developments and anticipates future research directions.

To provide in-depth examinations, each volume links experimental philosophy to a key philosophical area. They provide historical overviews alongside case studies, reviews of current problems and discussions of new directions. For upper-level undergraduates, postgraduates and professionals actively pursuing research in experimental philosophy these are essential resources.

Titles in the series include:

Advances in Experimental Epistemology, edited by James R. Beebe
Advances in Experimental Moral Psychology, edited by Hagop Sarkissian and Jennifer Cole Wright
Advances in Experimental Philosophy and Philosophical Methodology, edited by Jennifer Nado
Advances in Experimental Philosophy of Aesthetics, edited by Florian Cova and Sébastien Réhault
Advances in Experimental Philosophy of Language, edited by Jussi Haukioja
Advances in Experimental Philosophy of Mind, edited by Justin Sytsma
Advances in Religion, Cognitive Science, and Experimental Philosophy, edited by Helen De Cruz and Ryan Nichols
Experimental Metaphysics, edited by David Rose

Advances in Experimental Philosophy of Logic and Mathematics

Edited by
Andrew Aberdein and Matthew Inglis

BLOOMSBURY ACADEMIC
LONDON • NEW YORK • OXFORD • NEW DELHI • SYDNEY

BLOOMSBURY ACADEMIC
Bloomsbury Publishing Plc
50 Bedford Square, London, WC1B 3DP, UK
1385 Broadway, New York, NY 10018, USA
29 Earlsfort Terrace, Dublin 2, Ireland

BLOOMSBURY, BLOOMSBURY ACADEMIC and the Diana logo
are trademarks of Bloomsbury Publishing Plc

First published in Great Britain 2019
Paperback edition first published 2021

Copyright © Andrew Aberdein, Matthew Inglis and Contributors, 2019

Andrew Aberdein, Matthew Inglis have asserted their right under the Copyright, Designs and Patents Act, 1988, to be identified as Editors of this work.

Series design by Catherine Wood
Cover image © Dieter Leistner / Gallerystock

All rights reserved. No part of this publication may be reproduced or transmitted in any form or by any means, electronic or mechanical, including photocopying, recording, or any information storage or retrieval system, without prior permission in writing from the publishers.

Bloomsbury Publishing Plc does not have any control over, or responsibility for, any third-party websites referred to or in this book. All internet addresses given in this book were correct at the time of going to press. The author and publisher regret any inconvenience caused if addresses have changed or sites have ceased to exist, but can accept no responsibility for any such changes.

A catalogue record for this book is available from the British Library.

Library of Congress Cataloging-in-Publication Data
Names: Aberdein, Andrew, editor.
Title: Advances in experimental philosophy of logic and mathematics / edited by Andrew Aberdein and Matthew Inglis.
Description: New York: Bloomsbury Academic, 2019. |
Series: Advances in experimental philosophy |
Includes bibliographical references and index.
Identifiers: LCCN 2018051142 (print) | LCCN 2019011823 (ebook) |
ISBN 9781350039025 (epdf) | ISBN 9781350039032 (epub) |
ISBN 9781350039018 (hardback)
Subjects: LCSH: Logic, Symbolic and mathematical.
Classification: LCC QA9 (ebook) | LCC QA9.A3875 2019 (print) |
DDC 511.3–dc23
LC record available at https://lccn.loc.gov/2018051142

ISBN: HB: 978-1-3500-3901-8
PB: 978-1-3502-1795-9
ePDF: 978-1-3500-3902-5
eBook: 978-1-3500-3903-2

Series: Advances in Experimental Philosophy

Typeset by Deanta Global Publishing Services, Chennai, India

To find out more about our authors and books visit
www.bloomsbury.com and sign up for our newsletters.

Contents

Notes on Contributors   vi

1   Introduction   *Andrew Aberdein and Matthew Inglis*   1
2   Methodological Triangulation in Empirical Philosophy (of Mathematics)   *Benedikt Löwe and Bart Van Kerkhove*   15
3   Animal Cognition, Species Invariantism, and Mathematical Realism   *Helen De Cruz*   39
4   The Beauty (?) of Mathematical Proofs   *Catarina Dutilh Novaes*   63
5   Can a Picture Prove a Theorem? Using Empirical Methods to Investigate Visual Proofs by Induction   *Josephine Relaford-Doyle and Rafael Núñez*   95
6   An Empirical Study on the Admissibility of Graphical Inferences in Mathematical Proofs   *Keith Weber and Juan Pablo Mejía-Ramos*   123
7   Does Anyone Really Think That ⌜φ⌝ Is True If and Only If φ?   *Robert Barnard and Joseph Ulatowski*   145
8   New Foundations for Fuzzy Set Theory   *Igor Douven*   173
9   What Isn't Obvious about "Obvious": A Data-Driven Approach to Philosophy of Logic   *Moti Mizrahi*   201
10   Philosophy and the Psychology of Conditional Reasoning   *David Over and Nicole Cruz*   225
11   Folk Judgments about Conditional Excluded Middle   *Michael J. Shaffer and James R. Beebe*   251

Index   277

Notes on Contributors

Andrew Aberdein is Professor of Philosophy at Florida Institute of Technology where he has taught since 2003. He is the coauthor of *Rhetoric: The Art of Persuasion*, Wooden, 2015, and coeditor of *The Argument of Mathematics*, Springer, 2013. He is particularly interested in argumentation theory, and its use in understanding science, technology, and mathematics.

Robert Barnard is Professor of Philosophy at the University of Mississippi. He is the coeditor of *The Continuum Companion to Metaphysics*, Continuum, 2012. He works primarily in areas related to metaphysics and epistemology (especially theories of truth), and the history of analytic philosophy.

James R. Beebe is Professor of Philosophy, Director of the Experimental Epistemology Research Group, and Member of the Center for Cognitive Science at the University at Buffalo. In addition to working on skepticism and epistemic externalism in mainstream epistemology, most of his research in recent years has focused on the empirical investigation of folk epistemology (i.e., how individuals think about knowledge, rational justification, and evidence).

Nicole Cruz is a postdoctoral researcher at the Department of Psychological Sciences, Birkbeck, University of London. Her main research area is the psychology of reasoning.

Helen De Cruz is Senior Lecturer in Philosophy at Oxford Brookes University. She is the author of *Religious Disagreement*, Cambridge, 2018, coauthor of *A Natural History of Natural Theology*, MIT Press, 2015, and coeditor of *Advances in Religion, Cognitive Science, and Experimental Philosophy*, Bloomsbury, 2016. Her research investigates how we as human beings—embodied, socially and materially embedded, cognitively limited—acquire beliefs about subjects that seem far removed from our everyday experience, such as in mathematics, science, and theology.

Igor Douven is a cognitive scientist and formal epistemologist. He is a CNRS Research Professor at Sorbonne University, Paris. He is the author of

The Epistemology of Indicative Conditionals, Cambridge, 2016, and editor of *Lotteries, Knowledge, and Rational Belief*, Cambridge, forthcoming. He has published over 100 articles in journals including *Mind*, *Journal of Philosophy*, and *The Philosophical Review*.

Catarina Dutilh Novaes is Professor of Philosophy at the Free University (VU) in Amsterdam. She is also an external member of the Munich Center for Mathematical Philosophy and one of the editors-in-chief of *Synthese*. She is the author of *Formal Languages in Logic*, Cambridge, 2012, and *Formalizing Medieval Logical Theories*, Springer, 2007, and coeditor of *The Cambridge Companion to Medieval Logic*, Cambridge, 2016. Her main fields of research are history and philosophy of logic, philosophy of mathematics, and social epistemology.

Matthew Inglis is Professor of Mathematical Cognition at Loughborough University. From 2010 to 2015 he worked as a Royal Society Research Fellow, and in 2014 he was awarded the Selden Prize by the Mathematical Association of America. He is the coauthor of *Does Mathematical Study Develop Logical Thinking? Testing the Theory of Formal Discipline*, World Scientific, 2016, and *An Introduction to Mathematical Cognition*, Routledge, 2018. His research focuses on the cognitive processes involved in understanding mathematical ideas.

Benedikt Löwe is Professor of Mathematical Logic and Interdisciplinary Applications of Logic at Universität Hamburg, a member of the Institute for Logic, Language and Computation at Universiteit van Amsterdam, and Overseas Fellow of Churchill College, Cambridge. His research includes mathematical logic and empirical studies of mathematical practice. He is Secretary General of the Division for Logic, Methodology and Philosophy of Science and Technology of the International Union for History of Science and Technology (DLMPST/IUHPST) and a member of the Académie Internationale de Philosophie des Sciences (AIPS).

Juan Pablo Mejía-Ramos is Associate Professor of Mathematics Education at Rutgers University. He is mainly interested in mathematical argumentation and proof, particularly the ways in which university students and research-active mathematicians construct, read, and present arguments and proofs in mathematics.

Moti Mizrahi is Associate Professor of Philosophy at Florida Institute of Technology. He is the editor of *The Kuhnian Image of Science: Time for a Decisive*

Transformation?, Rowman and Littlefield, 2018. He is an associate editor of *Philosophia*. He has published extensively on the philosophy of science, the scientific realism/antirealism debate, the epistemology of philosophy, and argumentation. His work has appeared in journals such as *Argumentation, Erkenntnis, Philosophical Studies, Studies in History and Philosophy of Science*, and *Synthese*.

Rafael Núñez is Professor of Cognitive Science at the University of California, San Diego. He is the author or coauthor of several books including *Where Mathematics Comes From: How the Embodied Mind Brings Mathematics into Being*, Basic Books, 2000. His research investigates high level cognitive phenomena such as conceptual systems, abstraction, and inference mechanisms through various perspectives such as mathematical cognition, the empirical study of spontaneous gesture, and field research with the Aymara culture in the Andes.

David Over is Emeritus Professor of Psychology at Durham University. He is the coauthor of *Rationality and Reasoning*, Psychology Press, 1996 and *If*, Oxford, 2004, the editor of *Evolution and the Psychology of Thinking*, Psychology Press, 2003, and the coeditor of *The Thinking Mind*, Psychology Press, 2016. His principal research area is the psychology of reasoning.

Josephine Relaford-Doyle is a graduate student in Cognitive Science at the University of California, San Diego. Her research interests include mathematical thinking, communication in mathematics, abstract thinking, probability, and education.

Michael J. Shaffer is Professor of Philosophy at St. Cloud State University. He is the author of *Quasi-factive Belief and Knowledge-like States*, Lexington, forthcoming, and *Counterfactuals and Scientific Realism*, Palgrave MacMillan, 2012. He is the coeditor of *What Place for the A Priori?*, Open Court, 2011. He is primarily interested in epistemology, logic and the philosophy of science.

Joseph Ulatowski is Senior Lecturer and Convenor of Philosophy, and Director of the Experimental Philosophy Research Group at the University of Waikato. For 2016–2019, he is a member of the Veritas Research Centre of Yonsei University, South Korea. He is the author of *Commonsense Pluralism about Truth: An Empirical Defence*, Palgrave Macmillan, 2017. His main area of interest is the nature of truth.

Bart Van Kerkhove is Associate Professor of Philosophy at Vrije Universiteit Brussel and also teaches at Universiteit Hasselt. He is the editor of *New Perspectives on Mathematical Practices*, World Scientific, 2009, and the coeditor of several other collections. Much of his research focuses on the philosophy of mathematical practice.

Keith Weber is Professor of Mathematics Education at Rutgers University. His research interests are in the mathematical cognition of doing advanced mathematics. He is particularly interested in mathematical proof, including how mathematicians and mathematics majors present, read, understand, and evaluate proofs.

1

Introduction

Andrew Aberdein and Matthew Inglis

There has been very little overt discussion on the experimental philosophy of logic or mathematics. So it may be tempting to assume that application of the methods of experimental philosophy to logic or the philosophy of mathematics is impractical or unavailing. That this would be a mistake is exhibited by at least three trends in recent research: a renewed interest in historical antecedents of experimental philosophy in philosophical logic; a "practice turn" in the philosophies of mathematics and logic; and philosophical interest in a substantial body of work in adjacent disciplines, such as the psychology of reasoning and mathematics education. Before turning to the specific contribution that we hope this book will make, we will offer a snapshot of each trend and address how they intersect with some of the standard criticisms of experimental philosophy. First, although experimental philosophy is often thought of as a twenty-first-century phenomenon primarily focused on questions in ethics and epistemology, it has some important anticipations in earlier projects in the philosophy of logic. The most significant is the work of Arne Næss and the Oslo Group (Næss, 1938, 1959, 1982; Tönnessen, 1951). For instance, Ingemund Gullvåg argued that to understand the meaning of a word such as "truth," it was "hardly sufficient that a single person registers his own reactions to this or that sentence, or makes pronouncements based on intuitions" (Gullvåg, 1955: 343). Instead, the Oslo group argued, systematic empirical investigations were required. The connections between the "empirical semantics" developed by the Oslo group and experimental philosophy have now begun to be made explicit by historians of philosophy and further developed by a new generation of researchers (Murphy, 2014; Barnard and Ulatowski, 2016; Chapman, 2018). This productive connection between the empirical methods of two different generations is continued in Barnard and Ulatowski's chapter in the present volume, discussed in greater detail below.

Secondly, in recent decades there has been a "practice turn" in the philosophy of mathematics, focusing on how mathematical research is actually conducted, rather than on the search for foundations for mathematics (Van Kerkhove and Van Bendegem, 2007; Mancosu, 2008). This has naturally led to an interest in empirical data about mathematical practice, a program dubbed "Empirical Philosophy of Mathematics" by some of its practitioners (Buldt et al., 2008; Löwe et al., 2010; Pantsar, 2015). There are several distinct axes along which the connections between the philosophy of mathematical practice and empirical work have been drawn. A significant body of work applies cognitive science research on mathematical reasoning to philosophical questions (Pease et al., 2013). This includes work on the status of mathematical knowledge (Cappelletti and Giardino, 2007; Pantsar, 2014); on the symbol systems of mathematics (De Cruz and De Smedt, 2013; Dutilh Novaes, 2013; Marghetis and Núñez, 2013); and on the role of diagrams and visualization in mathematics (Giaquinto, 2007; Hamami and Mumma, 2013). Moreover, modern mathematicians increasingly employ online tools for collaboration. This produces a considerable amount of potential data for researchers interested in mathematical practice, giving rise to another strategy for the investigation of that practice (Martin and Pease, 2013; Martin, 2015; Pease et al., 2017). Although logical practice may have received less attention than its mathematical counterpart, some researchers in the philosophy of logic have pursued a practice turn of their own, modeled on that in the philosophy of mathematics (Dutilh Novaes, 2012). As with its sister program in philosophy of mathematics, advocates of the philosophy of logical practice stress that too much attention has been paid to foundational issues at the expense of philosophical questions that arise elsewhere, such as in the application of logic to artificial intelligence, game theory, linguistics, and other disciplines. For example, the burgeoning research program of "argumentation mining," which applies corpus-based techniques to extract and analyze arguments across large bodies of text, may be seen as a logical counterpart to the use of big data techniques in analysis of mathematical practice (Moens, 2018).

Thirdly, there is a growing awareness of how much research in adjacent disciplines has anticipated the research questions of experimental philosophy of logic and mathematics. Philosophers of logic have an extensive body of research on the psychology of reasoning to draw upon (Johnson-Laird, 2006). Lately, some work in the intersection of philosophy of logic and psychology of reasoning has made the relationship to experimental philosophy explicit (Pfeifer, 2012; Pfeifer and Douven, 2014; Ripley, 2016). There is also a substantial research tradition in mathematics education that addresses questions of immediate relevance to the philosophy of mathematical practice (Heinze, 2010; Weber et al., 2014; Weber and Mejía-Ramos, 2015; Alcock et al., 2016). Understanding mathematical practice is important for education researchers for at least two reasons. First, understanding the behavior of expert mathematicians helps to decide what the purpose of a mathematics curriculum should be. If a particular activity is highly valued in expert mathematical practice then this perhaps provides a reason for mathematics students to be exposed to some appropriate version of it (see, for example, Ball and Bass, 2000; Harel and Sowder, 2007; Lampert, 1990; Weber et al., 2014). Second, studying the in-the-moment strategies adopted by expert practitioners (in any domain) might provide suggestions for how to develop interventions that assist learners to develop expertise. An example of this approach can be found in the work of Alcock, Hodds, Roy, and Inglis (2015). They studied the reading behavior of research mathematicians, and used these insights to develop training materials that encouraged undergraduates to adopt similar strategies. These training materials significantly increased the amount students learned from reading a mathematical text.

In addition, there is now an emerging tradition of interdisciplinary work, applying quantitative techniques to address traditionally philosophical questions, such as mathematical aesthetics (Inglis and Aberdein, 2015, 2016). Some of this work has been presented as an enquiry into "mathematical cultures" (Löwe, 2016; Larvor, 2016). Likewise, Reuben Hersh, one of the forerunners of the practice turn, has lately called for "a unified, distinct scholarly activity of mathematics studies: the study of mathematical activity and behavior" (Hersh, 2017: 335). We regard the present volume as, in part, a contribution to the integrative work required for this project.

The advent of experimental philosophy has not been without controversy and has provoked a salutary debate on the proper methods of philosophical enquiry. One of the most prominent critiques is the "expertise defense" of traditional philosophical practice (Nado, 2014; Mizrahi, 2015). This maintains that surveys of nonphilosophers have limited bearing on the arguments of philosophers since, as experts, philosophers can be expected to be immune from the errors and biases exhibited by nonexperts. This debate has given rise to a substantial literature. However, the experimental philosophies of mathematics and logic seem to have ready responses to the expertise defense. Many studies of mathematical practice focus on professional mathematicians, placing the expertise of the participants essentially beyond dispute. Nonetheless, this is not universally true; for instance, some philosophers (e.g., De Cruz, 2016) have used results from the numerical cognition literature to draw conclusions about the ontology of natural numbers. Participants in numerical cognition studies include nonmathematical adults, children, and even nonhuman animals.

An important difference between mainstream experimental philosophy and work focused on mathematics is that studies in the latter tradition typically ask their participants—be they mathematicians, children, or animals—about mathematics, not about philosophy. (This is just as well, for in Hersh's famous formulation, "the typical working mathematician is a platonist on weekdays and a formalist on Sundays" (Hersh, 1979: 32). Such insouciance would not bode well for the resolution of philosophical dilemmas.) In this respect experimental philosophy of mathematics is similar to psychological work on reasoning relevant to debates in the philosophy of logic. Here too participants are typically drawn from a more general population. (Although there clearly is such a thing as logical expertise; for a start, people can be trained to be better at logical reasoning (Attridge et al., 2016).) Just as mathematicians/children/fish are asked about mathematics not philosophy, participants in reasoning studies are asked object-level questions about everyday reasoning, not specialized questions about logical hypotheses that might predict or explain such reasoning. On this basis David Ripley has argued that these studies are better placed to answer the expertise objection than studies relevant to debates in ethics or epistemology (Ripley, 2016).

Nonetheless, there is a substantial body of psychological research that reveals a divergence between best practice in reasoning (at least, as defined by logicians) and how lay people actually reason. There is also a substantial body of work critiquing these results. Broadly speaking, they lend themselves to four possible responses:

1. Lay people are to blame: they routinely make damaging errors in their inferential practices;
2. Psychologists are to blame: they fail to understand the relationship between formal and informal reasoning, and thereby design experiments which show only that good reasoners can be hoodwinked by artificial examples;
3. Logicians are to blame: they persist in defending systems of formal inference which do not describe the legitimate inferential practices of ordinary folk;
4. No one is to blame: logicians might well be right that formal logic is the best way to reason, but in many (perhaps most) real-world circumstances it takes too much cognitive effort to do so.

Many such studies, especially in early psychology of reasoning work, are presented as supporting the first response. However, they can often be reinterpreted in support of one of the others. In particular, much research of this sort is implicitly deductivist (and often classicist): it presumes that the best account of human inference will always be deductive logic (and often that classical logic is the best or only viable system of deductive logic). Hence such work is undermined by the successful modeling of informal, nondeductive patterns of inference in argumentation theory (Zenker, 2018) or nonclassical logics (Aberdein and Read, 2009). The moral may be that, as with many sciences, theoretical and empirical approaches should be mutually reinforcing: logicians need the empirical research conducted by psychologists of reasoning to corroborate their claim of faithfulness to actual reasoning; psychology of reasoning needs to be informed by current research in logic if it is to stay relevant.

Recent research suggests that even preverbal children can exhibit behavior consistent with logical reasoning (Cesana-Arlotti et al., 2018). Children as young as twelve months were presented with stimuli either complying with or violating simple inferential rules, such as disjunctive syllogism, $p \vee q, \neg p \vdash q$. That they looked longer at violating cases than they did at stimuli consistent with those rules, just as adults do, suggests that they found those cases incongruous. This provides an echo of a far older debate. The ancient logician Chrysippus argued that dogs employ disjunctive syllogism, since a scent hound, tracking a quarry to a crossroads and eliminating all but one of the exits, will (or so Chrysippus claims) immediately take the last exit without further checks of the trail. The story has been retold many times, with at least four different morals:

1. dogs use logic, so they are as clever as humans;
2. dogs use logic, so using logic is nothing special;
3. dogs reason well enough without logic;
4. dogs reason better for not having logic (for details, see Aberdein, 2008).

The third option may be closest to Chrysippus's own; it may also be the best take on the empirical research. That is, such studies do not attribute conscious, reflective awareness of any system of logic to dogs (or infants). Rather, they demonstrate that logic succeeds in tracking the pre-theoretical reasoning not just of the logically educated, but of pretty much anyone capable of rational thought.

It is sometimes argued that logic, as an *a priori* discipline, is immune from revisionary pressures that apply to natural science. If this is so, then there may be little room for empirical research in logic. On the other hand, there is a tradition, associated with W. V. O. Quine in particular, of treating logic as continuous with the natural sciences (Bryant, 2017). In recent years, this debate has been characterized in terms of "anti-exceptionalism" about logic (Hjortland, 2017; Read, 2019). However, we do not need to resolve the debate in order to observe that it is less damaging to our concerns than it may first appear. Even if one concedes that the truths of logic are analytic and necessary—that is, true in virtue of their meaning and such that they could not have been different—our knowledge of these truths is still fallible. So we may expect the methods whereby we come to know these truths to have much in common with the methods whereby we learn truths in the natural sciences, even though the truths of those disciplines are neither analytic nor necessary.

This collection is intended to consolidate and develop the three trends identified above: the reappraisal of the Oslo Group; the practice turn in the philosophies of logic and mathematics; and the reintegration into these philosophies of empirical work from adjacent disciplines. The ten chapters are divided equally between the philosophies of mathematics and logic. Their authors include some of the leading figures in each of the areas of research discussed above. Several chapters are methodological analyses of the applicability of empirical techniques to these areas of philosophy, but many (also) include actual empirical results. They demonstrate a wide variety of different empirical methods, including experiments, surveys, and data mining.

Benedikt Löwe and Bart Van Kerkhove's chapter, "Methodological triangulation in empirical philosophy of mathematics," is written by two of the leading figures in the philosophy of mathematical practice. They survey the uses that have been found for a variety of different empirical methods in philosophy, emphasizing that the experimental method in the strict sense is only one of them. They argue for methodological triangulation in empirical philosophy, that is, the employment of a battery of different empirical methods to compensate for the biases and limitations implicit in any one of them. Their chapter provides a helpful introduction to the potential that empirical methods offer for the philosopher of mathematics (or logic). In particular, they rehearse a sartorial analogy that Löwe has proposed elsewhere for the different levels of integration between philosophy and empirical methods (Löwe, 2016: 36). He distinguishes "ready-to-wear," the philosophical exploitation of existing, independently conducted empirical research, from "bespoke," which involves more direct collaboration, such as philosophers designing projects to be conducted by empirical researchers, and "do-it-yourself" (homespun?), in which the philosopher conducts all aspects of the research. (We might add that such cross-disciplinary work can cut both ways: empirical researchers can develop the interest and expertise necessary to address philosophical questions. Indeed, some philosophical questions, including many posed by the philosophies of mathematical and logical practice, are already within the remit of nearby empirical disciplines.) It is important to stress, as Löwe and Van Kerkhove do, that this is not a hierarchy of quality. If you are lucky enough to find off-the-peg clothes that are a good fit, they may be much better value than bespoke. And making your own clothes is unlikely to have a good outcome unless you acquire significant expertise. Likewise, when existing empirical studies address the right questions, ready-to-wear studies can be highly effective. The remaining chapters in this collection report on studies of all three varieties.

Helen De Cruz's work in the philosophy of mathematics has long made use of empirical results (De Cruz, 2006, 2016). Her chapter, "Animal cognition, species invariantism and mathematical realism," is a notable piece of ready-to-wear empirical philosophy. She uses a variety of results from numerical cognition (especially neurological and animal work) to tackle a recently influential argument in the philosophy of mathematics. The "evolutionary debunking" argument against moral realism suggests that our moral beliefs cannot be objectively true if they are the result of a highly contingent evolutionary process. If we had evolved from animals with very different social behavior (and there are many such species) then we would have a quite different set of moral intuitions, so why imagine that those intuitions track the truth? Mathematical realism, the view that our mathematical beliefs are objectively true, has obvious similarities to moral realism. So might there not be an analogous evolutionary debunking argument against mathematical realism too? However, De Cruz demonstrates that there is significant evidence that the mathematical behavior of animals is substantially convergent, which suggests that the analogy fails; if anything, the empirical data provide support for mathematical realism.

The next chapter, "The beauty (?) of mathematical proofs," also makes extensive use of existing empirical research. We have already noted Catarina Dutilh Novaes's research on logical practice; besides the history and philosophy of logic she also works on social epistemology and the philosophy of mathematics, as in this chapter. She coordinates a number of disparate literatures to propose a novel approach to the aesthetics of mathematical proof grounded in empirical work on affective responses to unexpectedness. The key idea is that in many situations mathematical judgments bring together epistemic and aesthetic components, and that we should not be surprised by this.

The next two chapters are drawn from the bespoke tradition: they report on original studies that were conducted by the authors to address (at least) philosophical questions. Both chapters look at aspects of visual reasoning in mathematics. This has been a controversial subject: an influential view maintains that visuals should play no role in mathematical proof, but a growing body of work suggests that this is an unrealistic, indeed harmful, idealization (Larvor, 2013, 2019). Josephine Relaford-Doyle and Rafael Núñez are both cognitive scientists—the latter is a coauthor of a landmark in the application of cognitive science to mathematics (Lakoff and Núñez, 2000). Their chapter, "Can a picture prove a theorem? Using empirical methods to investigate visual proofs by induction," reports an empirical study that investigates how undergraduate students with and without formal mathematical training use images to justify mathematical claims. They focus on visual induction proofs, and find results that challenge James Brown's (nonempirical) claim that such proofs are immediately understandable for people without mathematical training (Brown, 1997).

Keith Weber and Juan Pablo Mejía-Ramos are mathematics educators. In their chapter, "An empirical study on the admissibility of graphical inferences in mathematical proofs," they investigate the admissibility of graphical inferences in proofs in real analysis. They conclude that the type of graphical inference is important to consider when addressing their question. In particular, Weber and Mejía-Ramos find support for the importance of distinguishing between metrical and nonmetrical graphical inferences (Larvor, 2019). A metrical graphical inference is one that depends for its success on the measurements of angles, lengths, and so on, being precisely correct, whereas a nonmetrical graphical inference does not; that is, the latter sort of inference is unaffected by local deformations in the diagrams at issue.

Where the first five chapters focus primarily on the philosophy of mathematics, the remaining five concentrate on the philosophy of logic. In their chapter, "Does anyone really think that $\ulcorner\varphi\urcorner$ is true if and only if φ?," the philosophers Robert Barnard and Joseph Ulatowski link together Arne Næss's early empirical work, their own recent replications of some of these results, and the contemporary debate on deflationary accounts of truth. As well as noting this chapter's contribution to philosophical theory, given the ongoing replication crisis in psychology (Chambers, 2017), it is worth explicitly remarking upon and celebrating Barnard and Ulatowski's successful replication of Næss's early findings.

Igor Douven's research lies at the intersection of several fields, including formal epistemology and cognitive science. His chapter, "New foundations for fuzzy set theory," seeks to rehabilitate fuzzy set theory as an account of vagueness by grounding it in recent empirical work on conceptual spaces. Conceptual spaces were developed by the cognitive scientist Peter Gärdenfors as a geometrical framework for the qualitative comparison of concepts along multiple dimensions (Gärdenfors, 2000). Douven argues that seeing fuzzy membership as the distance of a point from a prototypical point in such a space is a productive approach to fuzzy set theory and, moreover, that there is empirical support for adopting this view from work conducted by cognitive psychologists.

The philosopher Moti Mizrahi's chapter, "What isn't obvious about 'obvious': A big data approach to philosophy of logic," uses a corpus linguistics approach to investigate the obviousness or otherwise of logic. Mizrahi reasons that if logic really was obvious, then the frequency with which logicians use the word "obvious" should correlate with deductive indicator words such as "necessary" and "certainly" but not with inductive indicator words such as "probably" or "likely." By analyzing a large corpus of text drawn from research papers published in logic, philosophy, mathematics, and biology journals, Mizrahi empirically tests these predictions. While he finds some support for the predictions, he also finds some results that require further explanation.

David Over and Nicole Cruz are both psychologists of reasoning. Their chapter, "Philosophy and the psychology of conditional reasoning," is a wide-ranging discussion of recent empirical work on conditional statements and its relationship to philosophy. The authors cite an extensive array of empirical studies to argue in favor of a Bayesian account of conditionals and against a mental model account. They conclude that much psychological research on conditionals has paid too little attention to philosophical and logical work. Remedying that oversight has already led to improved empirical studies, and promises to go further.

In their chapter, "Folk judgments about conditional excluded middle," the philosophers Michael J. Shaffer and James R. Beebe employ empirical studies to motivate a novel analysis of so-called Bizet/Verdi conditionals:

- If Bizet and Verdi had been compatriots, Bizet would have been Italian.
- If Bizet and Verdi had been compatriots, Verdi would have been French.

Across three experiments Shaffer and Beebe find evidence for Alchourrón et al.'s (1985) "belief revision" theory of counterfactuals, in line with the tradition of the Ramsey Test. Interestingly, they reject the alternative accounts from Lewis (1973) and Stalnaker (1981) by coordinating analyses from both quantitative and qualitative data.

The experimental philosophy of logic and mathematics has been quietly thriving for some time. We hope that this collection will form an indispensable resource for future research in the field.

References

Aberdein, A. (2008). Logic for dogs. In S. D. Hales (Ed.), *What Philosophy Can Tell You about Your Dog*, 167–81. Chicago, IL: Open Court.

Aberdein, A. and Read, S. (2009). The philosophy of alternative logics. In L. Haaparanta (Ed.), *The Development of Modern Logic*, 613–723. Oxford: Oxford University Press.

Alchourrón, C., Gärdenfors, P. and Makinson, D. (1985). On the logic of theory change: Partial meet functions for contraction and revision. *Journal of Symbolic Logic*, 50:510–30.

Alcock, L., Ansari, D., Batchelor, S., Bisson, M.-J., De Smedt, B., Gilmore, C., Göbel, S. M., Hannula-Sormunen, M., Hodgen, J., Inglis, M., Jones, I., Mazzocco, M., McNeil, N., Schneider, M., Simms, V. and Weber, K. (2016). Challenges in mathematical cognition: A collaboratively-derived research agenda. *Journal of Numerical Cognition*, 2(1):20–41.

Alcock, L., Hodds, M., Roy, S. and Inglis, M. (2015). Investigating and improving undergraduate proof comprehension. *Notices of the AMS*, 62(7):742–52.

Attridge, N., Aberdein, A. and Inglis, M. (2016). Does studying logic improve logical reasoning? In C. Csíkos, A. Rausch and J. Szitányi (Eds.), *Proceedings of the 40th Conference of the International Group for the Psychology of Mathematics Education, Szeged, Hungary, August 3–7, 2016*, volume 2, 27–34, Szeged: PME.

Ball, D. L. and Bass, H. (2000). Making believe: The collective construction of public mathematical knowledge in the elementary classroom. In D. Phillips (Ed.), *Yearbook of the National Society for the Study of Education, Constructivism in Education*, 193–224. Chicago, IL: University of Chicago Press.

Barnard, R. and Ulatowski, J. (2016). Tarski's 1944 polemical remarks and Næss's "experimental philosophy". *Erkenntnis*, 81(3):457–77.

Brown, J. R. (1997). Proofs and pictures. *British Journal for the Philosophy of Science*, 48:161–80.

Bryant, A. (2017). Resolving Quine's conflict: A Neo-Quinean view of the rational revisability of logic. *Australasian Journal of Logic*, 14(1):30–45.

Buldt, B., Löwe, B. and Müller, T. (2008). Towards a new epistemology of mathematics. *Erkenntnis*, 68:309–29.

Cappelletti, M. and Giardino, V. (2007). The cognitive basis of mathematical knowledge. In M. Leng, A. Paseau and M. Potter (Eds.), *Mathematical Knowledge*, 74–83. Oxford: Oxford University Press.

Cesana-Arlotti, N., Martín, A., Téglás, E., Vorobyova, L., Cetnarski, R. and Bonatti, L. L. (2018). Precursors of logical reasoning in preverbal human infants. *Science*, 359(6381):1263–66.

Chambers, C. D. (2017). *The Seven Deadly Sins of Psychology: A Manifesto for Reforming the Culture of Scientific Practice*. Princeton, NJ: Princeton University Press.

Chapman, S. (2018). The experimental and the empirical: Arne Næss's statistical approach to philosophy. *British Journal for the History of Philosophy*, 26(5): 961–81.

De Cruz, H. (2006). Towards a Darwinian approach to mathematics. *Foundations of Science*, 11:157–96.

De Cruz, H. (2016). Numerical cognition and mathematical realism. *Philosophers' Imprint*, 16(16):1–13.

De Cruz, H. and De Smedt, J. (2013). Mathematical symbols as epistemic actions. *Synthese*, 190(1):3–19.

Dutilh Novaes, C. (2012). Towards a practice-based philosophy of logic: Formal languages as a case study. *Philosophia Scientiæ*, 16(1):71–102.

Dutilh Novaes, C. (2013). Mathematical reasoning and external symbolic systems. *Logique & Analyse*, 221:45–65.

Gärdenfors, P. (2000). *Conceptual Spaces*. Cambridge, MA: MIT Press.

Giaquinto, M. (2007). *Visual Thinking in Mathematics: An Epistemological Study*. Oxford: Oxford University Press.

Gullvåg, I. (1955). Criteria of meaning and analysis of usage. *Synthese*, 9:341–61.

Hamami, Y. and Mumma, J. (2013). Prolegomena to a cognitive investigation of Euclidean diagrammatic reasoning. *Journal of Logic, Language and Information*, 22(4):421–48.

Harel, G. and Sowder, L. (2007). Towards a comprehensive perspective on proof. In F. Lester (Ed.), *Second Handbook of Research on Mathematics Teaching and Learning*, 805–42. Washington, DC: National Council of Teachers of Mathematics.

Heinze, A. (2010). Mathematicians' individual criteria for accepting theorems and proofs: An empirical approach. In G. Hanna, H. N. Jahnke and H. Pulte (Eds.), *Explanation and Proof in Mathematics: Philosophical and Educational Perspectives*, 101–11. Dordrecht, The Netherlands: Springer.

Hersh, R. (1979). Some proposals for reviving the philosophy of mathematics. *Advances in Mathematics*, 31:31–50.

Hersh, R. (2017). Mathematics as an empirical phenomenon, subject to modeling. *Journal of Indian Council of Philosophical Research*, 34(2):331–42.

Hjortland, O. T. (2017). Anti-exceptionalism about logic. *Philosophical Studies*, 174(3):631–58.

Inglis, M. and Aberdein, A. (2015). Beauty is not simplicity: An analysis of mathematicians' proof appraisals. *Philosophia Mathematica*, 23(1):87–109.

Inglis, M. and Aberdein, A. (2016). Diversity in proof appraisal. In B. Larvor (Ed.), *Mathematical Cultures: The London Meetings 2012-2014*, 163–79. Basel: Birkhäuser.

Johnson-Laird, P. N. (2006). *How We Reason*. Oxford: Oxford University Press.

Lakoff, G. and Núñez, R. E. (2000). *Where Mathematics Comes From: How the Embodied Mind Brings Mathematics into Being.* New York, NY: Basic Books.

Lampert, M. (1990). When the problem is not the question and the solution is not the answer: Mathematical knowing and teaching. *American Educational Research Journal,* 27(1):29–63.

Larvor, B. (2013). What philosophy of mathematical practice can teach argumentation theory about diagrams and pictures. In A. Aberdein and I. J. Dove (Eds.), *The Argument of Mathematics,* 239–53. Dordrecht, The Netherlands: Springer.

Larvor, B. (2016). What are mathematical cultures? In S. Ju, B. Löwe, T. Müller and Y. Xie (Eds.), *Cultures of Mathematics and Logic: Selected Papers from the Conference in Guangzhou, China, November 9–12, 2012,* 1–22. Basel: Birkhäuser.

Larvor, B. (2019). From Euclidean geometry to knots and nets. *Synthese,* Forthcoming.

Lewis, D. (1973). *Counterfactuals.* Cambridge, MA: Harvard University Press.

Löwe, B. (2016). Philosophy or not? The study of cultures and practices of mathematics. In S. Ju, B. Löwe, T. Müller and Y. Xie (Eds.), *Cultures of Mathematics and Logic: Selected Papers from the Conference in Guangzhou, China, November 9–12, 2012,* 23–42. Basel: Birkhäuser.

Löwe, B., Müller, T. and Müller-Hill, E. (2010). Mathematical knowledge as a case study in empirical philosophy of mathematics. In B. Van Kerkhove, J. P. Van Bendegem and J. De Vuyst (Eds.), *Philosophical Perspectives on Mathematical Practice,* 185–203. London: College Publications.

Mancosu, P. (Ed.) (2008). *The Philosophy of Mathematical Practice.* Oxford: Oxford University Press.

Marghetis, T. and Núñez, R. (2013). The motion behind the symbols: A vital role for dynamism in the conceptualization of limits and continuity in expert mathematics. *Topics in Cognitive Science,* 5(2):299–316.

Martin, U. (2015). Stumbling around in the dark: Lessons from everyday mathematics. In A. Felty and A. Middeldorp (Eds.), *CADE-25,* volume 9195 of *LNAI,* 29–51, Cham: Springer.

Martin, U. and Pease, A. (2013). Mathematical practice, crowdsourcing, and social machines. In J. Carette, D. Aspinall, C. Lange, P. Sojka and W. Windsteiger (Eds.), *CICM 2013,* volume 7961 of *LNAI,* 98–119. Berlin: Springer.

Mizrahi, M. (2015). Three arguments against the expertise defense. *Metaphilosophy,* 46(1):52–64.

Moens, M.-F. (2018). Argumentation mining: How can a machine acquire common sense and world knowledge? *Argument & Computation,* 9:1–14.

Murphy, T. S. (2014). Experimental philosophy: 1935–1965. *Oxford Studies in Experimental Philosophy,* 1:325–68.

Nado, J. (2014). Philosophical expertise. *Philosophy Compass,* 9(9):631–41.

Næss, A. (1938). Common sense and truth. *Theoria,* 4:39–58.

Næss, A. (1959). Logical equivalence, intentional isomorphism, and synonymity as studied by questionnaires. *Synthese,* 10:471–79.

Næss, A. (1982). A necessary component of logic: Empirical argumentation analysis. In E. M. Barth and J. L. Martens (Eds.), *Argumentation: Approaches to Theory Formation*, 9–22. Amsterdam: John Benjamins.

Pantsar, M. (2014). An empirically feasible approach to the epistemology of arithmetic. *Synthese*, 191(17):4201–29.

Pantsar, M. (2015). Assessing the "empirical philosophy of mathematics". *Discipline Filosofiche*, 25(1):111–30.

Pease, A., Guhe, M. and Smaill, A. (2013). Developments in research on mathematical practice and cognition. *Topics in Cognitive Science*, 5(2):224–30.

Pease, A., Lawrence, J., Budzynska, K., Corneli, J. and Reed, C. (2017). Lakatos-style collaborative mathematics through dialectical, structured and abstract argumentation. *Artificial Intelligence*, 246:181–219.

Pfeifer, N. (2012). Experiments on Aristotle's thesis: Towards an experimental philosophy of conditionals. *The Monist*, 95(2):223–40.

Pfeifer, N. and Douven, I. (2014). Formal epistemology and the new paradigm psychology of reasoning. *Review of Philosophy and Psychology*, 5:199–221.

Read, S. (2019). Anti-exceptionalism about logic. *Australasian Journal of Logic*, Forthcoming.

Ripley, D. (2016). Experimental philosophical logic. In W. Buckwalter and J. Sytsma (Eds.), *A Companion to Experimental Philosophy*, 523–34. Chichester: Wiley Blackwell.

Stalnaker, R. (1981). A defense of conditional excluded middle. In W. Harper, R. Stalnaker and G. Pearce (Eds.), *Ifs*, 87–104. Dordrecht, The Netherlands: D. Reidel.

Tönnessen, H. (1951). The fight against revelation in semantical studies. *Synthese*, 8(3–5):225–34.

Van Kerkhove, B. and Van Bendegem, J. P. (Eds.) (2007). *Perspectives on Mathematical Practices: Bringing Together Philosophy of Mathematics, Sociology of Mathematics, and Mathematics Education*. Dordrecht, The Netherlands: Springer.

Weber, K., Inglis, M. and Mejía-Ramos, J. P. (2014). How mathematicians obtain conviction: Implications for mathematics instruction and research on epistemic cognition. *Educational Psychologist*, 49(1):36–58.

Weber, K. and Mejía-Ramos, J. P. (2015). On relative and absolute conviction in mathematics. *For the Learning of Mathematics*, 35(2):15–21.

Zenker, F. (2018). Logic, reasoning, argumentation: Insights from the wild. *Logic and Logical Philosophy*, 27(4):421–51.

2

Methodological Triangulation in Empirical Philosophy (of Mathematics)

Benedikt Löwe and Bart Van Kerkhove

1 Introduction

This chapter is a reflection on using empirical methods in philosophy. Most of the issues discussed in this chapter are general and not restricted to any particular branch of philosophy; however, since both authors are philosophers of mathematics, our point of departure and guiding motivation will be the philosophy of mathematics, in particular the philosophy of mathematical practice. We shall observe that this research area has a number of methodological issues in common with *Experimental Philosophy*, explore these, and make a case for a *multiple-method approach*, also known as *methodological triangulation*.

In §§ 2 and 3, we give a brief overview of the two research areas *Philosophy of Mathematical Practice* and *Experimental Philosophy*, respectively. We shall observe that the name "experimental philosophy" suggests that the latter exclusively uses the experimental method rather than empirical methods in general; this observation leads us to a reflection on the term "experimental," issues of nomenclature, and the plurality of methods in empirical research in § 4. We highlight (in § 5) that the plurality of methods used in research is particularly important for empirical research in fields where the isolation and control of variables is either difficult or undesirable; in the social sciences, this paradigm is known, following Campbell and Fiske (1959), as *methodological triangulation*, and we discuss it in § 6. Finally, in § 7, we return to the philosophy of mathematical practice and give examples of the use of triangulation in our field.

2 Philosophy of mathematical practice

Philosophy of Mathematical Practice is an approach toward philosophy of mathematics that takes the practice of mathematicians into account. Its research community is motivated by the fact that the foundational debates in the philosophy of mathematics in the early twentieth century had resulted in a focus on a highly idealized version of mathematics, often ignoring the practices of working mathematicians.[1] The titles of volumes such as *What is Mathematics, Really?* (Hersh, 1997) or *Towards a Philosophy of Real Mathematics* (Corfield, 2003) express the frustration of the respective authors with a philosophy of mathematics dealing with a strongly sanitized version of the discipline. The conferences on *Perspectives on Mathematical Practices* in Brussels, organized by Jean Paul Van Bendegem and the second author, provided a venue for equally aggravated philosophers to meet with researchers from various empirical disciplines studying mathematical practice (cf. Van Kerkhove and Van Bendegem, 2007; Van Kerkhove, 2009; Van Kerkhove et al., 2010); Mancosu (2008) used the term *The Philosophy of Mathematical Practice* as the title for his edited volume that brings together authors "joined by the shared belief that attention to mathematical practice is a necessary condition for a renewal of the philosophy of mathematics" (Mancosu, 2008: 2). In 2009, the research community formed the *Association for the Philosophy of Mathematical Practice* (APMP; with Mancosu and Van Bendegem among the nine founding members), which states its aims and scope as follows:

> Over the last few years approaches to the philosophy of mathematics that focus on mathematical practice have been thriving. Such approaches include the study of a wide variety of issues concerned with the way mathematics is done, evaluated, and applied, and in addition, or in connection therewith, with historical episodes or traditions, applications, educational problems, cognitive questions, etc. We suggest using the label "philosophy of mathematical practice" as a general term for this gamut of approaches, open to interdisciplinary work. (APMP, 2017)

We observe that in this mission statement of the APMP, philosophy of mathematical practice is not primarily defined in terms of its object, but rather by its methods: it is a term for a "gamut of approaches" applied to the philosophy of mathematics. Löwe (2016: 31) reports that

> at the inaugural conference of the *Association for the Philosophy of Mathematical Practice* in Brussels there was a critical discussion of the term "Philosophy of

mathematical practice." Its syntactic form "philosophy of *X*" suggests that there is an object "mathematical practice" whose philosophy it is studying. . . . This view was in general rejected by the participants of the inaugural conference; instead, the consensus was that philosophy of mathematical practice is an approach (or a collection of approaches) to philosophy of mathematics.

Other terms than "philosophy of mathematical practice" have been used that make it more apparent that the research community is determined by an approach rather than a subject matter, among them "empirical philosophy of mathematics" (Löwe et al., 2010), "practice-based philosophy of mathematics" (Dutilh Novaes, 2012) or "(socio-)empirically informed philosophy of mathematics" (Müller-Hill, 2009, 2011).

The "gamut of approaches" invoked by the APMP includes historical approaches, sociological approaches, methods from cognitive science, and cognitive psychology, as well as mathematics education. The wide range of this list resulted in a cautionary commentary by Jullien and Soler (2014: 228; emphasis in the original) remarking that these are

> not, strictly speaking, approaches "in the philosophy of mathematical practice."
> . . . They are, rather, . . . *non*-philosophical perspectives on mathematical practice that are *used* by philosophers of mathematical practice or, more prudently, on which *some* philosophers of mathematical practice *can find* [it] *relevant* to rely.

The approaches used are methods of empirical social research (broadly speaking), which has two consequences. First, philosophers of mathematics are rarely trained in the methods of empirical social research. This lack of expertise requires either extensive training or increased collaboration with researchers from other disciplines (cf. Löwe, 2016: § 5). Second, since there is hardly any well-established methodology for using empirical results in philosophical arguments, reflection about the methods to be used is a worthwhile first step of this particular philosophical endeavor. We contribute to this reflective process in this chapter.[2]

3 Experimental philosophy

"Experimental philosophy is a new interdisciplinary field that uses methods normally associated with psychology to investigate questions normally associated with philosophy" (Knobe et al., 2012). This definition of the field of *Experimental Philosophy* establishes it as a field defined by its method rather than its subject, reminding us of the discussion in § 2. Another similarity is that experimental philosophy grew out of a methodological dissatisfaction with the *status quo*: experimental philosophers such as Knobe (2007: 72) and Prinz (2008: 199) have criticized traditional philosophers for making empirical claims on the basis of rather limited and potentially biased data;[3] they propose to replace expert intuition by the aforementioned "methods normally associated with psychology." While the method of experimental philosophy has been mostly applied in epistemology, philosophy of mind, philosophy of language, and ethics, in principle, there are no obstacles to applying these methods to any subfield of philosophy, including philosophy of mathematics. We observe that the mentioned definition of "experimental philosophy" does not mention the experimental method specifically (cf. § 4).[4]

Given these close similarities, it is not surprising that experimental philosophy had to deal with the two issues mentioned at the end of § 2: researchers in experimental philosophy had to develop skills in psychological research methods and the field had to reflect on its own methodology. Concerning the latter point, from its very inception, experimental philosophy was embroiled in a methodological debate with the armchair philosophers it aimed to criticize (cf., for example, Williamson, 2004; Sosa, 2007; Symons, 2008) and reacted by providing a manifesto (Knobe and Nichols, 2008) with a discussion of the method, its goals, and its limitations. Over the years, experimental philosophers have remained acutely aware and keenly interested in discussions of methods (e.g., Alexander, 2010; Sandis, 2010; Strickland and Suben, 2012; Sytsma and Livengood, 2012; Andow, 2016). As a reaction to the recent replication crisis (cf., for example, Begley and Ioannidis, 2015; Baker, 2016), the experimental philosophy community reacted by setting up the *XPhi Replicability Project* in order to reach a reliable estimate of the replicability of empirical results in experimental philosophy (the organizers of the project, Florian Cova and Brent Strickland, expect to publish its results in the journal *Review of Philosophy and Psychology*). As a consequence, experimental philosophy is a field with a high degree of methodological reflection and thus, we consider it beneficial for philosophers of mathematical practice to learn from the experiences made by experimental philosophers.

Let us return to the use of the term "experimental" in the name of the field. Until recently, the term *experimental philosophy* would have rather evoked the title of Margaret Cavendish's 1666 *Observations Upon Experimental Philosophy* in the minds of readers. Since in Cavendish's time, science was referred to as "natural philosophy," the term *experimental philosophy* here stands for the empirically based method of the natural sciences that only rather recently had embraced the controlled experiment as its most significant tool:

> The seventeenth century witnessed the rise of experiment as a means of acquiring knowledge about nature. It is not as if experiments had had no role in natural philosophy or medicine before the seventeenth century, but rather that the quantity, nature and significance of early modern experiments were markedly superior to what had preceded them in any period in the history of science. (Anstey, 2014: 103)

It is easy to forget that the experimental method which many consider paradigmatic for the scientific method as a whole is a relatively modern addition to the toolbox of the scientist. We use this as an excuse to explore the meaning of the word "experimental" in the subsequent section.

4 What is in a name?

According to Merriam-Webster (2017), an experiment is "an operation or procedure carried out under controlled conditions in order to discover an unknown effect or law, to test or establish a hypothesis, or to illustrate a known law." In general, the model for an experimental set-up has three types of variables: *independent variables* manipulated by the experimenter, *dependent variables* that are being measured in order to determine the effect that the change of the independent variables has, and *extraneous* or *controlled variables*. The treatment of these three types of variables constitutes the experimental method:

> The experimental method is defined by the manipulation of independent variables and the measurement of dependent variables. Extraneous variables are either controlled or allowed to vary randomly. In particular, care is taken to remove any variables that are confounded with the independent variables. (Healey and Proctor, 2003: ix)

The term *controlled variables* evokes the image of laboratory control in physical experiments where, for example, the temperature or the pressure during the experiment is kept constant, but this type of control is hardly ever possible in the social sciences where control of the extraneous variables is usually achieved statistically by random assignments of test subjects (or, more generally, units of analysis) to different conditions. In this chapter, we are using the term "control" to refer to both mentioned types of control.

In textbooks on scientific methodology, *experiment* is usually contrasted with *observation* which is missing the feature of manipulation of the independent variables and the control over the extraneous variables.[5] However, not all definitions of the word *experiment* emphasize all three mentioned aspects in order to contrast experiments and observations; in some cases, only the active manipulation of the independent variables is used as the crucial difference between an experiment and an observation: "To experiment is to isolate, prepare, and manipulate things in hopes of producing epistemically useful evidence" (Bogen, 2017: § 1). In the social science literature, studies that lack control of the extraneous variables via randomized assignments of test subjects to different conditions are often called *quasi-experimental* or *correlational*. It is also interesting to note that the stark contrast between experiment and observation did not always exist; for example, Anstey (2014: 105) reports that "in early modern English, the expression 'observation and experiment' . . . is often used as a hendiadys."

The usage of the words "experiment" and "experimental" are more multifaceted than any textbook definitions can capture. Let us add one particularly striking example of nomenclature from the experimental philosophy literature: Prinz (2008) proposed to demarcate the boundary line between *experimental* and *empirical* based on whether the empirical work is done by the philosophers themselves or rather by other scientists ("mining the data" vs "collecting the data"):

> Some philosophers make use of empirical results that have been acquired by professional scientists. . . . These results are used to support or refute philosophical theories. I will call this approach "empirical philosophy." Other philosophers also conduct their own psychological experiments, an approach known as "experimental philosophy." (Prinz, 2008: 196)

Löwe (2016: § 4.3) argues that Prinz's demarcation line according to the person collecting the data is important, but the terms "experimental" and "empirical" for the corresponding types of philosophy are infelicitous.[6]

The confusion about when to properly call a method "experimental" is related to the fact that the idea of a well-defined universal notion of the scientific method is a gross oversimplification.[7] Hoyningen-Huene (2013: 4sq) reports that

> belief in the existence of scientific methods conceived of as strict rules of procedure has eroded. . . . Research situations . . . are so immensely different from each other across the whole range of the sciences and across time that it appears utterly impossible to come up with one set of universally valid methodological rules.

It is not fruitful to get into a nomenclatural debate about the correct definition of the term "experimental." Instead, it is more apt to realize that the diairetic distinctions discussed are important features that allow us to describe particular methodological approaches and discuss their strengths and weaknesses. The distinctions that we have seen so far are:

1. Does the researcher actively manipulate the environment or not?
2. Does the researcher control the features that are currently not under investigation or not?
3. Is the philosopher actively involved either in the design of the research or even in the actual empirical research itself or not?

Of course, these three mentioned distinctions are not the only ones that matter. No discussion of methodological distinctions of empirical work is complete without mentioning the divide between *qualitative* and *quantitative* research. The language used in our original discussion of the experimental method (in particular, the use of the word "variable") suggests that an experimental set-up presupposes a quantitative methodology, but this is not the case: questions of whether the researcher should actively manipulate the environment or not feature very prominently in discussions of qualitative research as well.

We summarize: there is a multitude of scientific methods, many ways to classify them in order to separate one method from another, each with its own advantages and disadvantages. We should like to propose the term *empirical philosophy* for the philosophical methodology that takes empirical data, produced according to one of the established empirical scientific methods, into account.[8] Both philosophy of mathematical practice and experimental philosophy discussed in §§ 2 and 3 fall under this label.

The philosophical methodology of incorporating empirical findings in philosophical arguments is underdeveloped; the methodological reflection of experimental philosophers mentioned in § 3 has started this necessary debate (cf. also Crupi and Hartmann, 2010).

5 "Medicine is not an exact science"

In § 4, we discussed a plethora of empirical methods, of which the experimental method in the strict sense, that is, involving manipulation of the environment under controlled or randomized conditions, is but one. So, it is striking that the dominant image of scientific activity is that of

hypothesis / experimental design / experiment / validation or refutation which entirely ignores the methodological variety of scientific practice. The overemphasis of the experimental method in popular accounts of the scientific method stems from the fact that in (parts of) the exact sciences, this particular methodology goes a long way.[9] Lieberson and Lynn (2002) lamented that the perceived equivalence of scientific rigor with the methods of the exact sciences is very harmful for the scientific methodology in the social sciences:

> Our thesis is that deeply ingrained in sociology and other social sciences is a special model of natural science that is exceptionally inappropriate. It is derived from physics, particularly the classical physics that existed before the beginning of the twentieth century.... The use of physics as the ideal model for sociology is so embedded in our thinking that the influence and appropriateness of this particular model is rarely questioned.... Our thesis is that other natural sciences actually offer epistemological and procedural models that are more relevant for the obstacles encountered in sociology and other social sciences. (Lieberson and Lynn, 2002: 2)

This complaint echoes the debates of the *Methodenstreit* of the late nineteenth century: in the *Methodenstreit*, the historical school of economics, led by Gustav von Schmoller (1838–1917), claimed that economic laws could not be independent of historical developments, and that economics should above all proceed on historical-empirical (inductive) grounds; in contrast, Carl Menger (1840–1921) argued that economics should adhere to the hypothetico-deductive methodology of the exact sciences.[10] In some respects, the *Methodenstreit* has never seen a truce.

A similar contrast with the method of the exact sciences is present in the oft-uttered adage "Medicine is not an exact science." The saying originated in the late nineteenth century in discussions of medical and forensic experts warning against overconfidence in the precision and definiteness of the statements of medical forensics.[11] The intended connotation of the sentence "Medicine is not an exact science" is closely related to our earlier discussion of the experimental method: as discussed in § 4, the main ingredients of the experimental method in the strict sense are manipulation of the independent variables and control of the extraneous variables: this usually requires an experimental setting in a laboratory, in which the experimenter can "isolate, prepare, and manipulate." Even in the exact sciences, it is questionable whether a researcher can truly control all relevant parameters.[12] But the field of (clinical) medicine usually deals with an individual living human being, some of whose variables certainly cannot and some should not be experimentally controlled and no random variation is possible. As a consequence, in the terminology used in § 4, medical practice is observational rather than experimental. When medical research is experimental, that is, done under controlled conditions in a laboratory *in vitro*, the applicability of these research results often suffers from the fact that they were obtained in controlled laboratory environments, but applied to human beings *in vivo*, embedded in a social environment. This phenomenon is witnessed by contradictory newspaper headlines of the type "Red wine is good for your health" and "Red wine is not good for you after all."

But it is not just that controlling variables is harder and possibly unethical in the medical sciences; also the concepts that medicine is dealing with are noticeably different from the concepts in physics. For example, the concept of "good for you" is normative, and therefore, its relevant standards are clearly determined by social, societal, and psychological norms in a way that concepts in the exact sciences usually are not. Notice that we do not take a position on whether the concepts in the natural sciences are socially constructed.[13] But while in the natural sciences, it is a respectable philosophical position to claim that the concepts of *force* or *gravity* are entirely independent of human culture and society, it would be bizarre to claim that a normative concept such as "good for you" could be defined in full generality independently of references to human psychology and society. When moving from medical sciences to the social sciences, this becomes even more patently obvious.

The crucial step in empirical social research to deal with the social dependence of the concepts is that of *operationalization*: the researcher models the socially dependent concept with an operationalized concept for which—within the confines of the given study—it can be determined with a reasonable degree of certainty whether the concept applies or not. In other words, the concepts we are dealing with are either *subjective*, *vague*, or *cluster concepts*[14] and operationalization replaces these concepts with a precisely defined concept.

Let us consider an example: if a political scientist aims to study the behavior of *typical Conservative Party voters*, she will need to operationalize this concept in order to have a precise assignment of the label *typical Conservative Party voter* to individual test subjects. For example, based on the past voting records, she counts a test subject as *typical Conservative Party voter* if they voted for the Conservative Party at least three times out of five. Note that this is not a definition of a *typical Conservative Party voter*: the researcher is very much aware that, for example, there might be typical Conservative Party voters who, for whatever reason, did not vote for the Conservative Party in the last few elections, creating a false negative (or similarly for a false positive). But as there is no precise definition, its role has to be played by the operationalized concept. This, in turn, requires arguments that this particular operationalization (producing false positives and negatives) does not affect the results of this study. The choice of operationalization and the arguments for it are confined to the concrete study that the political scientist is working on.[15]

In the context of the philosophical debate, the need to operationalize is closely connected to the distinction between data and phenomena highlighted by Bogen and Woodward (1988: 305*sq*):

> Data, which play the role of evidence for the existence of phenomena, for the most part can be straightforwardly observed. However, data typically cannot be predicted or systematically explained by theory. By contrast, well-developed scientific theories do predict and explain facts about phenomena. Phenomena are detected through the use of data, but in most cases are not observable in any interesting sense of that term.

In our example, while the data, that is, the voting record and the individual behavior, can be straightforwardly observed, the phenomenon that the political scientist aims to study cannot be read off directly from the data, but requires conceptual consolidation into a phenomenology.[16] The replacement of the concept from the original research question with the operationalized concept is one of the underlying reasons for criticism of the use of empirical techniques in philosophy: for example, Kauppinen (2007: 97) argues that experimental studies about the concept of knowledge only study the *folk concept* of knowledge which may not be related to the philosophical concept that—as philosophers—we aim to study. As said before, the onus is on the empirical researcher to argue that the replacement of the informal concept from the original question with the operationalized concept is adequate within the context of the study.

All of the above issues combined result in a situation where the empirical social scientists need to remain aware that any question they are asking is not universal, but restricted to a particular cultural and social context, and that data collected from arbitrary sources is most likely meaningless. Kauppinen (2007: 101*sq*) reminds us that

> it should be obvious that when philosophers appeal to "us" in making their claims, the extension is limited to those who are competent with the concept in question. After all, what incompetent users of a concept say about a given case does not tell us anything about the concept we are interested in—someone who has no relevant pre-theoretical knowledge about the concept cannot manifest

it. Nobody would test a Gettier analysis by asking a small child whether the person in the case described knows or not, or count the child's response as a counterexample. And children are only the most obvious example. On many theories of concept possession, competence with a concept is a matter of degree and context. This is to deny that there is, strictly speaking, such a thing as a "competent speaker of English," for example.

It is important to note that in practice, determining whether a given test subject has the competence to answer our question is not isolated from and independent of the answer they give; this raises the specter of circular reasoning and leads to meta-phenomena such as Collins's *experimenter's regress* (Collins, 1981). Empirical social research is fundamentally hermeneutic in nature: traditionally, hermeneutics is the skill of interpreting *oratio obscura* by taking commonsense reasoning, the context, and the background of the speaker into account; it is governed by principles such as presumptions and *aequitas hermeneutica* (Scholz, 1999: § I.7–10). Thus, the hermeneutic approach is in conflict with some of the fundamental principles of the experimental approach which requires the control of external factors, whereas the hermeneutic principles require the *consideration* of the effects of many of these external factors as an important part of the interpretation. In Dilthey's terms, the social sciences are about *Verstehen* rather than *Erklären*.[17]

Taken together, the impossibility of controlling variables and the ubiquity of cluster concepts that need to be replaced by an operationalization in the context of empirical studies requires an approach to the empirical social sciences that includes multiple viewpoints, angles, and methods. Tilly (2004: 597) summarizes the situation as follows:

> A common prejudice, to be sure, divides the social world into phenomena that are suitable for quantification (population distributions, social mobility, etc.) and those that are irreducibly qualitative: conversation, narratives, biography, ethnography, and history often serve as examples. Formalisms clearly can and do apply, however, to these phenomena as well. . . . Formalisms blindly followed induce blindness. Intelligently adopted, however, they improve vision. Being obliged to spell out the argument, to check its logical implications, and to examine whether the evidence conforms to the argument promotes both visual acuity and intellectual responsibility.

The meta-method used for obtaining this "visual acuity and intellectual responsibility" is that of *methodological triangulation* which we shall discuss in the next section.

6 Triangulation

The method of *triangulation*, sometimes called *mixed method research*, is a (meta-)methodology in social research. The use of the term can be traced to a technical paper focusing on different quantitative methods by Campbell and Fiske (1959).[18] The seminal book (Webb et al., 1966: 3) generalized the idea to the general notion of combination of multiple methods of data collection:

> The most persuasive evidence comes through a triangulation of measurement processes. If a proposition can survive the onslaught of a series of imperfect measures, with all their irrelevant error, confidence should be placed in it. Of course, this confidence is increased by minimizing error in each instrument and by a reasonable belief in the different and divergent effects of the sources of error.

This meta-method of combining methodologies in the study of the same phenomenon was taken up by Denzin (1970, 1978) in his influential textbook.[19]

Triangulation allows the empirical researcher to avoid the danger of circularity implicit in the hermeneutic approach: for example, if determining the competence of a test subject and the subject's answers to the questionnaire are closely related and depend on each other, then a solution is to use one method to determine competence of the test subject and an entirely different method to do the survey; or, multiple different operationalizations of an informal concept could be used in order to provide an argument that the answer does not depend on the operationalization.[20] Empirical results obtain stability if several methodological approaches converge on them and only in this way can social phenomena be sufficiently stabilized in order to draw reliable conclusions. Following Geertz (1973), the results of such multimodal approaches are often called *thick descriptions*.[21]

Even though triangulation has been an accepted meta-method in the social sciences for decades, a large amount of social empirical research is done using a single method, often an experimental method. It has been argued that this state of affairs is one of the reasons why the social sciences, and in particular psychology, have been hit especially hard by the replication crisis (Stroebe and Strack, 2014; Earp and Trafimow, 2015); one of the proposed solutions to the replication crisis is a form of triangulation (cf. Schooler, 2014, there called *meta-science*).[22]

We propose to embrace and apply the meta-methodology of triangulation for what we called *empirical philosophy* in § 4. We should hasten to add that this proposal does not mean that we think that any empirical approach that uses but one method is flawed and without merit. In fact, as can be expected from the standard popular account of falsificationalist science, the experimental approach is particularly good at producing negative results and casting doubt on hypothesis and premises. This critical approach to hidden premises is very much needed in philosophy; for example, the results of Weinberg et al. (2001), without any further triangulation by other methods, are sufficient to cast doubt on the assumption that epistemic intuitions are universal traits of human beings independent of which culture they grew up in: consequently, if a philosophical position is based on this assumption, the results require any philosophers who holds these views to reevaluate their position.

7 Triangulation in philosophy of mathematical practice

After our excursion into general methodology or philosophy of science, we now return to the philosophy of mathematical practice: we apply the lessons learned from considering methodological issues in the empirical social sciences. As in the general case in § 6, we emphasize that not all empirically based philosophical analyses that rely on a single method are problematic: the experimental method is very good at pointing out hidden assumptions about homogeneity or universality and casting doubt on these. For example, Núñez (2011) uses an experimental (or quasi-experimental) approach and without any further triangulation by other methods, his results show convincingly that there are problems with the assumption that previous experiments prove that the cognitive number line is innate and physically represented in the human brain.

However, raising issues with assumptions is not all that philosophy of mathematics wants to do, and in analogy to the use of triangulation in the empirical social sciences, we propose to embrace methodological triangulation as a meta-method in empirical philosophy of mathematics and close our chapter by giving three examples of good triangulation practice in philosophy of mathematics:

Our first example is the study on knowledge attributions among mathematicians. Müller-Hill (2009) discusses a questionnaire study in the tradition of experimental philosophy: test subjects from the target group (research mathematicians, defined as people who claim to have research or university teaching experience in mathematics) were given questionnaires with fictitious stories about the production of mathematical knowledge and the resulting data were statistically evaluated (cf. also Löwe et al., 2010). In one of the stories, a fictitious mathematician believes in the truth of a theorem (the proof of the *Jones conjecture*) he published, but finds out during a talk he attends that there is a counterexample to his result. 61.3 percent of the test subjects answer positively to the question "Does John know that the Jones conjecture is false?" and yet 71 percent of the test subjects answer positively to the question "Did John know that the Jones conjecture was true on the morning before the talk?." As in the methodological debates about experimental philosophy, traditional philosophers of mathematics have criticized questionnaire-based studies like this with arguments similar to the objections listed in Knobe and Nichols (2008): in the terminology used in § 5, traditional philosophers observe that such a study replaces the notion of mathematical knowledge with an operationalized concept that is based on actual knowledge utterances by research mathematicians and claim that this operationalized concept is quite different from the philosophical concept of mathematical knowledge.

Müller-Hill (2011) then interpreted the quantitative results of the initial study by means of frequency and cluster analysis and supplements this with the text comments provided by the test subjects (Müller-Hill, 2011: § 3.2.2). The qualitative processing of the various answers in the initial study allowed Müller-Hill to isolate a number of different answer profiles for each fictitious scenario. These then formed the basis of a number of in-depth interviews, serving as a qualitative complement to the initial study and thus an additional triangulation point (Müller-Hill, 2011: § 4).

The second example is studies by the research groups at Loughborough University and Rutgers University of the conviction of mathematicians about the correctness of a mathematical result. The degree of conviction about mathematical correctness is one of those highly context-dependent concepts that defy precise definitions and thus need to be operationalized. In a preliminary study, Inglis and Mejía-Ramos (2008) provided a typology of responses to the question about the level of persuasion by a proof. Weber (2008) and Weber and Mejía-Ramos (2011) conducted interview studies on goals and methods of reading proofs by research mathematicians; these qualitative studies generated hypotheses that were then tested with a survey study with 118 research-active mathematicians by Mejía-Ramos and Weber (2014). Further focusing on the mechanisms of reading proofs, an eye-tracking study of mathematicians reading proofs by Inglis and Alcock (2012) produced data that was provided to Weber and Mejía-Ramos (2013) who then further analyzed the data with other methods; their study was followed up by a reply of Inglis and Alcock (2013).

This example is of particular interest since it illustrates how triangulation can result in what Stegenga (2012) calls "discordant evidence": on some particular issues relevant to the study, the two research groups reached different conclusions using different methods. On the basis of introspective data gathered through interviews and questionnaires, Weber and Mejía-Ramos claimed that mathematicians typically skim mathematical texts before reading them in detail, but the eye-tracking study by Inglis and Alcock suggested the opposite. Such instances of scientific dissent fueled by methodological pluralism are an opportunity and invitation to renewed and intensified discussion, also on the methodological level, that one would hope leads eventually to a more nuanced understanding of the phenomena.

Our final example is the study of the mathematical peer review process: the original epistemological motivation of Geist et al. (2010) was the question about mathematical certainty in the case of knowledge by testimony via published proofs. Geist et al. (2010) studied the role of the mathematical peer review process in ascertaining correctness of proofs in the published literature, in particular, whether the fact that a paper went through the mathematical peer review process serves as a warrant for mathematical correctness. The idealized description of the peer review process would assume that every referee meticulously checks the mathematical details of proofs, but that idealized description does not match with the experiences of many if not most mathematicians. In the absence of empirical studies of the mathematical peer review process, Geist et al. (2010) systematically assess the process via a (small-scale) qualitative questionnaire study of editors of mathematical journals and a study of referee agreement comparing mathematical conferences with conferences from other subjects. Other researchers arrived at similar empirical questions from very different theoretical backgrounds: both Weintraub and Gayer (2001) and Heinze (2010) are sociologically interested in the criteria used by the community of mathematicians to decide whether a new proof becomes accepted; Weintraub and Gayer study this via an in-depth analysis of one particular case (the proof of the existence of a competitive equilibrium by Arrow and Debreu) and Heinze investigates this by means of a questionnaire study. Recently, the sociologist Greiffenhagen (personal communication) started using interviews with journal editors and case studies dealing with similar questions.

The last example is different from the other two in that it is not the same research group approaching the question from different angles, but different unrelated groups, in this case from three different disciplines (philosophy, mathematics education, and sociology), coalesce to the same questions from very different starting points with possibly incompatible motivations. Triangulation in the philosophy of mathematics, that is, approaching philosophical questions empirically from multiple angles and with multiple methods should not ultimately rely on serendipitous convergence of interests: in order to increase the number of triangulated studies in empirical philosophy of mathematics, more dialogue among researchers is needed to overcome this motivational incommensurabilities and find common ground (as argued by Löwe, 2016: 39). We hope that this chapter will convince some of our colleagues to commit some of their energy to this worthwhile effort.

Notes

1. For a more detailed discussion of this research community and its relationship to philosophy, cf. (Löwe, 2016: §§ 2 & 3).
2. We should like to emphasize that in this chapter, we do not intend to contribute to the discussion of whether empirical methods are fruitful in philosophy or the even more overarching question about the aims of philosophy. We discuss the methodological issues under the tacit assumption that we have decided to apply empirical methods in philosophy.
3. Inglis and Aberdein (2016: 166) introduce the term "examplar philosophers" for those who "offer an example . . ., assert that [it] has a given property, and appeal to the readers' intuition for agreement."
4. Since this chapter is not chiefly about experimental philosophy, we shall not be able to do justice to the scope and diversity of the experimental philosophy literature that includes much more than what we can discuss in this section.
5. Cf., for example, (Oehlert, 2010: 2): "We are in control of experiments, and having that control allows us to make stronger inferences about the nature of differences that we see in the experiment. This . . . distinguishes an experiment from an observational study. An observational study also has treatments, units, and responses. However, in the observational study we merely observe which units are in which treatment groups; we don't get to control that assignment."
6. Using a sartorial analogy, Löwe (2016: § 4.3) introduces the terms *ready-to-wear* or *off-the-rack* for the use of empirical results in philosophy where the philosopher had no input in the design of the empirical study, *bespoke* for a project in which the philosopher works very closely with the empirical scientist and designs an experiment or other observational activity jointly with her, and *do-it-yourself* as the extreme case of bespoke where the philosopher becomes an empirical scientist and does the empirical work herself.
7. Cf. (Andersen and Hepburn, 2016).
8. As a label, *empirical philosophy* has often been identified with the Vienna Circle conception of what philosophy ought to be like: subordinate to (natural) science

(Benjamin, 1939). The label was also used in the edited volume (Wagenknecht et al., 2015) where the focus is mostly on qualitative methods and in their *Prolegomena to an Empirical Philosophy of Science*, Osbeck and Nersessian (2015: 15) explicitly contrast their approach with experimental philosophy: "An example is evident in relation to the recent trend of adopting empirical methods from psychology to inform philosophy, including philosophy of science. This 'experimental philosophy' is something of a curiosity, because nowhere has the ambiguity of 'empirical' created more problems than in the discipline of psychology."

We emphatically do not intend to include these connotations in our notion of empirical philosophy. An alternative label that could be used to avoid the confusion with other uses of the term is *empirically based philosophy*.

9 We take the liberty of not defining the term "exact sciences," not because we like to stay vague and suggestive, but because this would open yet another subtle terminological debate of which this chapter already has its fair share. If the reader is vexed by this, she or he can replace the term "exact sciences" with "classical physics" throughout. We should, however, like to emphasize that even in the prototypical exact science, physics, not all theories lend themselves to experimental testing in the proper sense: for example, you cannot do controlled experiments concerning planetary motions.

10 For details, cf. (Huff, 1984: 28–34) and (Schulak and Unterköfler, 2011: 21–28).

11 Cf. (Mnookin, 2007). Recently, a new wave of overconfidence of this type was dubbed the *CSI effect* and studied in (Shelton et al., 2006; Cole and Dioso-Villa, 2009; Smith et al., 2011; Cole, 2015).

12 We will not go into details here and refer the reader to the rich literature on *ceteris paribus* clauses in philosophy of science (cf., for example, Earman et al., 2003; Schrenk, 2007).

13 We refer the reader to the relevant literature (e.g., Longino, 2016).

14 That is, concepts defined by a list of criteria none of which are necessary or sufficient, akin to the Wittgensteinian *Familienähnlichkeit* (Gasking, 1960; Cooper, 1972; Parsons, 1973).

15 Note that this distinguishes operationalization from Carnapian *explication*.

16 Löwe and Müller (2011) discuss this in the framework of *conceptual modeling* and use the distinction to shed some light on the methodological discussion between the experimental philosophers and the armchair philosophers. They note that the instability of introspective intuitions has been criticized long before the advent of experimental philosophy: "The intuitive findings of different people, even of different experts, are often inconsistent. . . . If agreement about usage cannot be reached within so restricted a sample as the class of Oxford Professors of Philosophy, what are the prospects when the sample is enlarged?" Mates (1958: 165).

17 Cf. also the discussion of the *That's-Not-All-There-Is Objection* against experimental philosophy (Knobe and Nichols, 2008: 10).

18 In (Campbell and Fiske, 1959), the term refers to actual triangles in tables of data. Nowadays, we usually think of the triangulation metaphor as mimicking geometrical triangulation in navigation, allowing to determine the location of an object by giving its coordinates from two different angles. This notion of triangulation is not to be confused with the notion used by Davidson (1982) which uses the same metaphor in a similar situation.

19 Denzin (1978) also gives a classification of types of triangulation: data triangulation, investigator triangulation (involving multiple researchers to avoid biases brought in by the researcher), theory triangulation (using multiple theoretical schemes), and methodological triangulation. We focus on methodological triangulation for the present discussion without claiming that the other types of triangulation are unimportant.

Green and Wortham (2018: 69–72) argue that W. E. B. Du Bois (1868–1963), the first African American to receive a doctorate from Harvard University, was using methodological triangulation much earlier than this, and give examples from studies performed by Du Bois and published between 1898 and 1904: "Throughout his research, Du Bois used a diverse array of research methods to gather data. . . . Few, if any sociologists of the era, were as imbued with a methodological conscience as the intrepid Du Bois Green and Wortham (2018)" (Heesen et al. 20xx).

Du Bois's cautious and critical attitude toward the research methods is best described by his notion of *diffidence* laid out in (Du Bois, 1899: 2–3): "The best available methods of sociological research are at present so liable to inaccuracies that the careful student discloses the results of individual research with diffidence; he knows that they are liable to error from the seemingly ineradicable faults of the statistical method, to even greater error from the methods of general observation, and, above all, he must ever tremble lest some personal bias, some moral conviction or some unconscious trend of thought due to previous training, has to a degree distorted the picture in his view."

20 In Denzin's terminology of Endnote 19, the former example would be methodological triangulation, the latter would be theory triangulation.

21 It is interesting to note that Collins's own solution to the experimenter's regress problem is a form of unintentional triangulation; in his view, "The regress is eventually broken by negotiation within the appropriate scientific community, a process driven by factors such as the career, social, and cognitive interests of the scientists, and the perceived utility for future work (Franklin and Perovic, 2016: § 1.2.1)."

22 Heesen et al. (20xx) propose a formal model to evaluate and discuss the merits of triangulation based on Du Bois's *diffidence* (cf. Endnote 19): they discuss criticism of methodological triangulation in the literature (e.g., that it produces "discordant evidence") and conclude that their model "vindicate[s the] ... use of methodological triangulation (Heesen et al., 20xx: § 4)."

References

Alexander, J. (2010). Is experimental philosophy philosophically significant? *Philosophical Psychology*, 23(3):377–89.

Andersen, H. and Hepburn, B. (2016). Scientific method. In E. N. Zalta (Ed.), *The Stanford Encyclopedia of Philosophy*. Summer 2016 edition.

Andow, J. (2016). Qualitative tools and experimental philosophy. *Philosophical Psychology*, 29(8):1128–41.

Anstey, P. R. (2014). Philosophy of experiment in early modern England: The case of Bacon, Boyle and Hooke. *Early Science and Medicine*, 19:103–32.

APMP (2017). Association for the Philosophy of Mathematical Practice. About. Website accessed on October 4, 2017.

Baker, M. (2016). 1,500 scientists lift the lid on reproducibility. *Nature*, 533(7604):452–54.

Begley, C. G. and Ioannidis, J. P. (2015). Reproducibility in science: Improving the standard for basic and preclinical research. *Circulation Research*, 116(1):116–26.

Benjamin, A. C. (1939). What is empirical philosophy? *Journal of Philosophy*, 36(19):517–25.

Bogen, J. (2017). Theory and observation in science. In E. N. Zalta (Ed.), *The Stanford Encyclopedia of Philosophy*. Summer 2017 edition.

Bogen, J. and Woodward, J. (1988). Saving the phenomena. *Philosophical Review*, 97(3):303–52.

Campbell, D. and Fiske, D. (1959). Convergent and discriminant validation by the multitrait-multimethod matrix. *Psychological Bulletin*, 56(2):81–105.

Cole, S. A. (2015). A surfeit of science: The "CSI effect" and the media appropriation of the public understanding of science. *Public Understanding of Science*, 24(2):130–46.

Cole, S. A. and Dioso-Villa, R. (2009). Investigating the "CSI effect" effect: Media and litigation crisis in criminal law. *Stanford Law Review*, 61(6):1335–73.

Collins, H. M. (1981). 'Son of seven sexes', the social destruction of a physical phenomenon. *Social Studies of Science*, 11(1):33–62.

Cooper, D. E. (1972). Definitions and 'clusters'. *Mind*, 81(324):495–503.

Corfield, D. (2003). *Towards a Philosophy of Real Mathematics*. Cambridge University Press.

Crupi, V. and Hartmann, S. (2010). Formal and empirical methods in philosophy of science. In F. Stadler, D. Dieks, W. J. González, S. Hartmann, T. Uebel and M. Weber (Eds.), *The Present Situation in the Philosophy of Science*, 87–98. Springer-Verlag.

Davidson, D. (1982). Rational animals. *Dialectica*, 36:318–27.

Denzin, N. K. (1970). *The Research Act: A Theoretical Introduction to Sociological Methods*, first edition. Aldine.

Denzin, N. K. (1978). *The Research Act: A Theoretical Introduction to Sociological Methods*, second edition. McGraw-Hill.

Du Bois, W. E. B. (1899). *The Philadelphia Negro: A Social Study*. University of Pennsylvania Press.

Dutilh Novaes, C. (2012). Towards a practice-based philosophy of logic: Formal languages as a case study. *Philosophia Scientiae*, 16:71–102.

Earman, J., Glymour, C. and Mitchell, S. (Eds.) (2003). *Ceteris Paribus Laws*. Springer-Verlag.

Earp, B. D. and Trafimow, D. (2015). Replication, falsification, and the crisis of confidence in social psychology. *Frontiers in Psychology*, 6:Article 621.

Franklin, A. and Perovic, S. (2016). Experiment in physics. In E. Zalta (Ed.), *The Stanford Encyclopedia of Philosophy*. Winter 2016 edition.

Gasking, D. (1960). Clusters. *Australasian Journal of Philosophy*, 38(1):1–36.

Geertz, C. (1973). Thick description: Toward an interpretive theory of culture. In C. Geertz (Ed.), *The Interpretation of Cultures. Selected Essays*, 3–30. Basic Books.

Geist, C., Löwe, B. and Van Kerkhove, B. (2010). Peer review and knowledge by testimony in mathematics. In B. Löwe and T. Müller (Eds.), *PhiMSAMP, Philosophy of Mathematics: Sociological Aspects and Mathematical Practice*, 155–78. College Publications.

Green, D. S. and Wortham, R. A. (2018). The sociological insight of W. E. B. Du Bois. *Sociological Inquiry*, 88(1):56–78.

Healey, A. F. and Proctor, R. W. (Eds.) (2003). *Handbook of Psychology. Volume 4: Experimental Psychology*. John Wiley & Sons.

Heesen, R., Kofi Bright, L. and Zucker, A. (20xx). Vindicating methodological triangulation. To appear in *Synthese*.

Heinze, A. (2010). Mathematicians' individual criteria for accepting theorems and proofs: An empirical approach. In G. Hanna, H. N. Jahnke and H. Pulte (Eds.), *Explanation and Proof in Mathematics. Philosophical and Educational Perspectives*, 101–11. Springer-Verlag.

Hersh, R. (1997). *What Is Mathematics, Really?* Oxford University Press.

Hoyningen-Huene, P. (2013). *Systematicity. The Nature of Science*. Oxford University Press.

Huff, T. E. (1984). *Max Weber and the Methodology of the Social Sciences*. Transaction Books.

Inglis, M. and Aberdein, A. (2016). Diversity in proof appraisal. In B. Larvor (Ed.), *Mathematical Cultures. The London Meetings 2012–2014*, Trends in the History of Science, 163–79. Birkhäuser.

Inglis, M. and Alcock, L. (2012). Expert and novice approaches to reading mathematical proofs. *Journal for Research in Mathematics Education*, 43(4):358–90.

Inglis, M. and Alcock, L. (2013). Skimming: A response to Weber and Mejía-Ramos. *Journal for Research in Mathematics Education*, 44(2):471–74.

Inglis, M. and Mejía-Ramos, J. P. (2008). How persuaded are you? A typology of responses. *Research in Mathematics Education*, 10:119–33.

Jullien, C. and Soler, L. (2014). Conceptions of mathematical practices: Some remarks. Commentary on "The impact of the philosophy of mathematical practice on the philosophy of mathematics", by Jean Paul Van Bendegem. In L. Soler, S. Zwart,

M. Lynch and V. Israel-Jost (Eds.), *Science after the Practice Turn in the Philosophy, History, and Social Studies of Science*, 227–37. Routledge.

Kauppinen, A. (2007). The rise and fall of experimental philosophy. *Philosophical Explorations*, 10(2):95–118.

Knobe, J. (2007). Experimental philosophy. *The Philosophers' Magazine*, 50:72–73.

Knobe, J., Buckwalter, W., Nichols, S., Robbins, P., Sarkissian, H. and Sommers, T. (2012). Experimental philosophy. *Annual Review of Psychology*, 63:81–99.

Knobe, J. and Nichols, S. (2008). An experimental philosophy manifesto. In J. Knobe and S. Nichols (Eds.), *Experimental Philosophy*, 3–16. Oxford University Press.

Lieberson, S. and Lynn, F. B. (2002). Barking up the wrong branch: Scientific alternatives to the current model of sociological science. *Annual Review of Sociology*, 28:1–19.

Longino, H. (2016). The social dimensions of scientific knowledge. In E. N. Zalta (Ed.), *The Stanford Encyclopedia of Philosophy*. Spring 2016 edition.

Löwe, B. (2016). Philosophy or not? The study of cultures and practices of mathematics. Selected papers from the conference in Guangzhou, China, 9–12 November 2012. In S. Ju, B. Löwe, T. Müller and Y. Xie (Eds.), *Cultures of Mathematics and Logic*, 23–42. Birkhäuser.

Löwe, B. and Müller, T. (2011). Data and phenomena in conceptual modelling. *Synthese*, 182(1):131–48.

Löwe, B., Müller, T. and Müller-Hill, E. (2010). Mathematical knowledge: A case study in empirical philosophy of mathematics. In B. Van Kerkhove, J. De Vuyst and J. P. Van Bendegem (Eds.), *Philosophical Perspectives on Mathematical Practice*, 185–203. College Publications.

Mancosu, P. (Ed.) (2008). *The Philosophy of Mathematical Practice*. Oxford University Press.

Mates, B. (1958). On the verification of statements about ordinary language. *Inquiry*, 1:161–71.

Mejía-Ramos, J. P. and Weber, K. (2014). How and why mathematicians read proofs: Further evidence from a survey study. *Educational Studies in Mathematics*, 85:161–73.

Merriam-Webster (2017). Experiment. In *Online Dictionary*. Website accessed on October 4, 2017.

Mnookin, J. L. (2007). Idealizing science and demonizing experts: An intellectual history of expert evidence. *Villanova Law Review*, 52(4):763–801.

Müller-Hill, E. (2009). Formalizability and knowledge ascriptions in mathematical practice. *Philosophia Scientiae*, 13(2):21–43.

Müller-Hill, E. (2011). *Die epistemische Rolle formalisierbarer mathematischer Beweise—Formalisierbarkeitsorientierte Konzeptionen mathematischen Wissens und mathematischer Rechtfertigung innerhalb einer sozio-empirisch informierten Erkenntnistheorie der Mathematik*. PhD thesis, Rheinische Friedrich-Wilhelms-Universität Bonn.

Núñez, R. (2011). No innate number line in the human brain. *Journal of Cross-Cultural Psychology*, 45(4):651–68.

Oehlert, G. W. (2010). *A First Course in Design and Analysis of Experiments.* Self-published.

Osbeck, L. M. and Nersessian, N. J. (2015). Prolegomena to an empirical philosophy of science. In S. Wagenknecht, N. J. Nersessian and H. Andersen (Eds.), *Empirical Philosophy of Science: Introducing Qualitative Methods into Philosophy of Science*, 13–35. Springer-Verlag.

Parsons, K. P. (1973). Three concepts of clusters. *Philosophy and Phenomenological Research*, 33(4):514–23.

Prinz, J. J. (2008). Empirical philosophy and experimental philosophy. In J. Knobe and S. Nichols (Eds.), *Experimental Philosophy*, 198–208. Oxford University Press.

Sandis, C. (2010). The experimental turn and ordinary language. *Essays in Philosophy*, 11(2):181–96.

Scholz, O. R. (1999). *Verstehen und Rationalität. Untersuchungen zu den Grundlagen von Hermeneutik und Sprachphilosophie.* Vittorio Klostermann.

Schooler, J. W. (2014). Metascience could rescue the "replication crisis." *Nature*, 515(7525):9.

Schrenk, M. (2007). *The Metaphysics of Ceteris Paribus Laws*, volume 16 of *Philosophische Analyse*. Ontos.

Schulak, E. M. and Unterköfler, H. (2011). *The Austrian School of Economics. A History of Its Ideas, Ambassadors, and Institutions.* Ludwig von Mises Institute.

Shelton, D. E., Kim, Y. S. and Barak, G. (2006). A study of juror expectations and demands concerning scientific evidence: Does the "CSI effect" exist? *Vanderbilt Journal of Entertainment and Technology Law*, 9:331–68.

Smith, S. M., Stinson, V. and Patry, M. W. (2011). Fact or fiction? The myth and reality of the CSI effect. *Court Review*, 47:4–7.

Sosa, E. (2007). Experimental philosophy and philosophical intuition. *Philosophical Studies*, 132:99–107.

Stegenga, J. (2012). Rerum concordia discors: Robustness and discordant multimodal evidence. In L. Soler, E. Trizio, T. Nickles and W. Wimsatt (Eds.), *Characterizing the Robustness of Science: After the Practice Turn in Philosophy of Science.* Springer-Verlag.

Strickland, B. and Suben, A. (2012). Experimenter philosophy: The problem of experimenter bias in experimental philosophy. *Review of Philosophy and Psychology*, 3:457–67.

Stroebe, W. and Strack, F. (2014). The alleged crisis and the illusion of exact replication. *Perspectives on Psychological Science*, 9(1):59–71.

Symons, J. (2008). Intuition and philosophical methodology. *Axiomathes*, 18:67–89.

Sytsma, J. and Livengood, J. (2012). Experimental philosophy and philosophical disputes. *Essays in Philosophy*, 13(1):145–60.

Tilly, C. (2004). Observations of social processes and their formal representations. *Sociological Theory*, 22(4):595–602.

Van Kerkhove, B. (Ed.) (2009). *New Perspectives on Mathematical Practices. Essays in Philosophy and History of Mathematics. Brussels, Belgium, March 26–28, 2007*. World Scientific.

Van Kerkhove, B., De Vuyst, J. and Van Bendegem, J. P. (Eds.) (2010). *Philosophical Perspectives on Mathematical Practice*. College Publications.

Van Kerkhove, B. and Van Bendegem, J. P. (Eds.) (2007). *Perspectives on Mathematical Practices. Bringing Together Philosophy of Mathematics, Sociology of Mathematics, and Mathematics Education*. Springer-Verlag.

Wagenknecht, S., Nersessian, N. J. and Andersen, H. (Eds.) (2015). *Empirical Philosophy of Science: Introducing Qualitative Methods into Philosophy of Science*. Springer-Verlag.

Webb, E. J., Campbell, D. T., Schwartz, R. D. and Sechrest, L. (1966). *Unobtrusive Measures. Nonreactive Research in the Social Sciences*. Rand McNally & Company.

Weber, K. (2008). How mathematicians determine if an argument is a valid proof. *Journal for Research in Mathematics Education*, 39:431–59.

Weber, K. and Mejía-Ramos, J. P. (2011). How and why mathematicians read proofs: An exploratory study. *Educational Studies in Mathematics*, 76:329–44.

Weber, K. and Mejía-Ramos, J. (2013). On mathematicians' proof skimming. A reply to Inglis and Alcock. *Journal for Research in Mathematics Education*, 44(2):464–71.

Weinberg, J. M., Nichols, S. and Stich, S. (2001). Normativity and epistemic intuitions. *Philosophical Topics*, 29:429–60.

Weintraub, E. R. and Gayer, T. (2001). Equilibrium proofmaking. *Journal of the History of Economic Thought*, 23(4):421–42.

Williamson, T. (2004). Philosophical 'intuitions' and scepticism about judgment. *Dialectica*, 58(1):109–53.

3

Animal Cognition, Species Invariantism, and Mathematical Realism

Helen De Cruz

1 Introduction

What can we infer from numerical cognition about mathematical realism? In this chapter, I will consider one aspect of numerical cognition that has received little attention in the literature: the remarkable similarities of numerical cognitive capacities across many animal species. This Invariantism in Numerical Cognition (INC) indicates that mathematics and morality are disanalogous in an important respect: proto-moral beliefs differ substantially between animal species, whereas proto-mathematical beliefs (at least in the animals studied) seem to show more similarities. This makes moral beliefs more susceptible to a contingency challenge from evolution compared to mathematical beliefs, and indicates that mathematical beliefs might be less vulnerable to evolutionary debunking arguments. I will then examine to what extent INC can be used to flesh out a positive case for mathematical realism. Finally, I will review two forms of mathematical realism that are promising in the light of the evolutionary evidence about numerical cognition, *ante rem* structuralism and Millean empiricism.

2 The contingency challenge

Moral realism is the view that moral claims, such as "slavery is wrong" or "Jane is a good person," are about facts and that we know some of these facts. Moral facts are normative: they not only describe what is the case (e.g., slavery is wrong, or giving people their freedom is right), but also what ought to be the case (e.g., people should never be enslaved). Such facts are different from natural facts (e.g., that water is composed of H_2O), but that does not make them any less true in the eyes of the moral realist. By contrast, moral antirealists contend that there are no moral facts. Traditionally, moral antirealists have argued that moral claims do not describe beliefs, but emotions (e.g., violence makes me feel bad; slavery makes me feel sorry for enslaved people). More recently, authors such as Sharon Street (2006, 2008) and Richard Joyce (2006) have argued against moral realism on evolutionary grounds. They worry that human moral intuitions and their resulting judgments are influenced by the peculiar evolutionary history of our species. Arriving at the correct moral beliefs, given the contingency of human evolution, would be a formidable and inexplicable instance of luck. It would be as if one set sail in the hope that the winds and tides will get one to Bermuda (Street, 2006: 121). As Street writes,

> There is a striking coincidence between the normative judgments we human beings think are true, and the normative judgments that evolutionary forces pushed us in the direction of making. I claim that the realist about normativity owes us an explanation of this striking fact, but has none. (Street, 2008: 207)

Street (2006) lists some moral concerns that are similar across many species, such as that survival is good, or that an obligation to care for one's offspring is greater than the obligation to help complete strangers. However, many other moral concerns are the result of the peculiar quirks of human evolution. For example, humans believe that helping unrelated strangers is a good thing, or that one should punish group members who do not follow social norms (Henrich et al., 2006). Such (proto)moral sentiments are not present in other primates (see e.g., Silk and House, 2011).

In the *Descent of Man* (1871) Darwin investigated, among many other topics, the evolution of the moral sense in humans. The overall project of that book was to show that although the difference between human cognitive capacities and those of others was substantial, including the human sense of morality, beauty, and religion, it was only a difference in degree and not in kind. Darwin sought to establish precursors of the moral sense in other animals. He conjectured that the evolution of moral capacities became unavoidable in cognitively complex social animals. As he wrote, "Any animal whatever, endowed with well-marked social instincts, the parental and filial affections being here included, would inevitably acquire a moral sense or conscience, as soon as its intellectual powers had become as well, or nearly as well developed, as in man" (Darwin, 1871: 71–72). But he also thought that their moral beliefs would vary depending on the social context in which members of the species would evolve. To illustrate this point vividly, he offered the following thought experiment: if humans had evolved from animals with a eusocial structure, our moral beliefs would be very different from the ones we currently hold. For instance, "our unmarried females would, like the worker-bees, think it a sacred duty to kill their brothers, and mothers would strive to kill their fertile daughters; and no one would think of interfering" (Darwin, 1871: 73). This is because in eusocial societies the long-term survival of the group trumps concerns of individual workers. Eusociality has evolved three to eleven times independently in nature, in various clades including insects, shrimps, and mammals (West and Gardner, 2010). If there are rational creatures on other planets the kinds of actions we think are morally reprehensible could be obligatory for them and vice versa.

A moral realist could respond to this challenge by arguing that in eusocial structures, nests rather than individual workers are the right-bearers, so the difference may not be so vast after all. One could also argue that morality is a uniquely human domain, which does not even arise in other animals. Other animals have proto-morality at best: dispositions that lead them to helping behavior, or that lead them to prefer prosocial over antisocial individuals, but no explicit moral norms governing social interactions. Still, Darwin's bee thought experiment stresses the contingency of moral beliefs upon our specific evolutionary history. Our moral beliefs are not just the product of evolution; they are the peculiar outcome of human evolution, a haphazard process that has favored unique social structures and behaviors.

Lillehammer (2010: 365) terms challenges of this kind the *contingency challenge*: "We would have had very different beliefs if certain things about us had been different, even supposing the relevant ethical facts to remain the same." A related, but distinct, challenge is the *inflexibility challenge*, which says that we would have had the same beliefs, even if the relevant facts had been different. Street's (2006) Darwinian dilemma for moral realists stresses the inflexibility of our moral beliefs: even if pain were morally good in some realist sense, we would still be inclined to disvalue it because evolution through natural selection leads us to disvalue pain, as it decreases survival and reproductive success. Both kinds of challenge are part of a broader kind of purported failure, which spells bad news for moral realism, the *tracking failure* (Lillehammer, 2010). The evolutionary challenge to ethics does not amount to the claim that moral beliefs are likely not truth-tracking because they are the outcome of a long, evolutionary process (this is probably the case for all our beliefs, and thus would render the challenge trivial). Rather, it is the more specific claim that moral beliefs are not truth-tracking because they depend on contingent facts about our evolutionary history.

In this chapter, I will not be concerned with the evolutionary debunking literature on moral realism, but rather, with the question of whether evolutionary challenges to moral realism could be extended to the mathematical domain (see also, for example, Clarke-Doane, 2012, 2014; De Cruz, 2016). This fits in a broader literature of the so-called "companions in guilt" arguments where relevant features of moral realism are argued to occur in other domains (e.g., logic, perception, mathematics) (see, for example, Rowland, 2016). The basic outline of such an argument holds that these two domains fall or stand together: if a challenge to moral realism proves fatal, it will also be fatal for that other domain. The question here is whether evolutionary debunking arguments, if successful against moral realism, also damage mathematical realism. Mathematical realism—analogous to moral realism—is the claim that mathematical statements such as "2 + 2 = 4" are about facts. In order to assert this, the mathematical realist posits that mathematical entities (e.g., the natural numbers), relations, or structures exist. A dominant position in mathematical realism is platonism, which holds that mathematical objects exist. They are abstract entities that exist independently of human minds, cultural constructs, language, and symbols. Clarke-Doane (2014, manuscript) has connected the contingency challenge, normally applied to the question of moral realism, to the Benacerraf-Field challenge to mathematical realism. Benacerraf (1973) originally formulated the following objection to mathematical platonism: if mathematical objects are outside of space-time, how can we establish a causal link between these objects and the minds of mathematicians? How can the physical brain of the mathematician get access to these remote mathematical facts? Field (1989) reformulated the challenge in more general terms. Field's formulation does not require a causal link (which is a controversial requirement in any case), but hinges on the fact that mathematical realists cannot explain the reliability of mathematical beliefs:

> We start out by assuming the existence of mathematical entities that obey the standard mathematical theories; we grant also that there may be positive reasons for believing in those entities . . . Benacerraf's challenge . . . is to . . . explain how our beliefs about these remote entities can so well reflect the facts about them . . . . If it appears in principle impossible to explain this, then that tends to undermine the belief in mathematical entities, despite whatever reason we might have for believing in them. (Field, 1989: 26)

How we should cash out "explain the reliability" is not easily resolved. Clarke-Doane (manuscript) argues that it should be spelled out in terms of safety: "In order to 'explain the reliability'" of our mathematical beliefs it is necessary to show that we could not have easily had false ones (using the method that we actually used to form them), even if, had we, they would have been false." This formulation responds both to the contingency challenge: our mathematical beliefs are not easily false (for instance, because they are not contingent upon our evolutionary history in a way that is pernicious), and to the inflexibility challenge: our mathematical beliefs are not inflexible. They track something that is independent from them, and that could be false.

Clarke-Doane (2016) argued that mathematical realists might not face an access worry because mathematical beliefs are arguably safe. This might be because our "core" mathematical beliefs could be evolutionarily inevitable. However, in a more recent paper (manuscript) Clarke-Doane argues that it would be very hard for a mathematical realist to show that mathematical beliefs, say, about set theory, are safe, because it would be difficult to show that "we could not have easily believed different axioms of set theory." Indeed, the fact that mathematicians disagree about many core claims in every mathematical area shows that mathematical beliefs do not meet this criterion. For instance, Edward Nelson rejected the successor axiom (every natural number has a successor). Because mathematical beliefs are presumably less colored by irrelevant influences such as religion and cultural background, Clarke-Doane (2014) thinks that this puts mathematical realists in a worse position than moral realists, as the latter can at least explain away moral disagreements as a result of distorting irrelevant factors, but the former cannot.

This interpretation of the Benacerraf-Field challenge places the bar for realists quite high, some might argue, impossibly high. Moreover, some epistemologists have argued that safety is not a useful criterion for knowledge (e.g., Bogardus, 2014). Nevertheless, spelling out the Benacerraf-Field challenge in terms of safety can be useful if we consider the evolutionary origins of mathematical beliefs, in particular, numerical beliefs. In this chapter, I will not be concerned with set theory or other mathematical propositions, but with the evolutionary basis of our ability to form beliefs about numbers at all.

There is a large literature that supports the view that formal mathematics depends on evolved capacities to deal with number, which is collectively sometimes referred to as "the number sense" (e.g., Dehaene, 2011). Two capacities are hypothesized to underlie animals' ability to deal with numbers: the object file system (OFS) which might underlie our ability to enumerate and keep in working memory small collections of items (up to three or four) precisely, a capacity that is called *subitizing*, and the approximate number system (ANS), which may underpin our capacity to estimate and compare larger collections (Feigenson et al., 2004).

Many cognitive scientists hold that the OFS and ANS lie at the basis of our ability to engage in more formal arithmetic abilities. Elizabeth Spelke (e.g., Spelke and Kinzler, 2007) has argued that ANS supplemented with language allows for the ability to engage in formal arithmetic. She finds support for the role of language in studies indicating that people who speak languages without exact number words cannot perform basic calculations exactly (e.g., 6 − 2 = 4), but their approximate numerical cognition is on par with numerate adults (Pica et al., 2004). Although it is limited in that it only allows for approximate numerical calculations, the ANS already allows for abstract numerical representations across modalities: preschool children can add and compare arrays of dots and sound sequences (Barth et al., 2005). Carey (2009) sees the OFS at the root of more formal arithmetical capacities. Her bootstrapping account emphasizes the role of subitizing in children's ability to learn the successor function in arithmetic. Children learn to associate the meanings of the first words in a count list (in English, "one," "two," and "three") with collections of one, two, and three items, which they can subitize. This explains why children tend to learn the meanings of number words in the same order: they first become one-knowers, then two-knowers, next three-knowers, and very occasionally, four-knowers. But because subitizing stops at three or four, they need to make an inductive generalization to learn the next words in the counting sequence. According to Carey, children then make the following induction: if "x" is followed by "y" in the counting sequence, adding an individual to a set with cardinal value x results in a set with cardinal value y.

The idea that these two capacities play a critical role in our ability to engage in formal arithmetic is not universally accepted (see for example, Rips et al., 2006, Rips et al., 2008). Some authors have argued that nonnumerical sensory properties, such as visual density and circumference, can explain the animal data and have questioned the existence of the ANS (e.g., Gebuis et al., 2016). That being said, the ANS and the OFS are still the predominant theories to explain animal numerical cognition. Authors such as Lourenco et al. (2012) have argued that people's ability to engage in approximate arithmetic (both symbolic and nonsymbolic) correlates with their mathematical abilities. This has been confirmed in a recent meta-analysis, although the correlation between mathematical skills and symbolic numerical abilities is stronger (Schneider et al., 2017). If formal arithmetic is dependent (in some causal or psychological sense) on the evolved number sense, it becomes relevant for mathematical realists to explore it in more detail. In particular, explaining the reliability of our mathematical abilities will involve reference to the number sense and the way it forms beliefs about magnitudes.

3 Invariantism in numerical cognition

Numerical cognition is a well-researched domain of higher cognition. While obviously it is not identical across species (for one thing, humans use Arabic numerals whereas mosquito fish do not), I will here examine striking similarities between the numerical capacities of animals from a wide variety of species and clades, which I will call INC. I will here look at four features of numerical cognition across species to argue the case for INC: numerical cognition is present in many different animal species, including in animals with small, simple nervous systems such as insects and spiders (3.1), it plays a crucial role in animal adaptive decision making (3.2), it shows similarities in computational characteristics and limitations across species (3.3) and, to the extent that it has been investigated, there is evidence that numerical cognition is the result of convergent cognitive evolution rather than common descent (3.4).

3.1 Numerical competence is present in a wide variety of clades

Most research on numerical competence has been conducted with primates, including rhesus monkeys, capuchin monkeys, and chimpanzees. For example, rhesus monkeys are able to order collections of items from 1 to 9 (Brannon and Terrace, 1998). Other mammals, including brown bears and dogs, are also capable of discriminating numerosities. For example, brown bears were trained to select among two screens the display that had the largest number of dots (even if sometimes that meant the overall lowest surface area, because the dots were smaller), using food reinforcements (Vonk and Beran, 2012). Domestic dogs were tested using a violation-of-expectation paradigm, where they saw simple calculations including "1 + 1 = 2," "1 + 1 = 1," and "1 + 1 = 3." Dogs looked longer at the incorrect outcomes, which is interpreted as showing they did not expect the incorrect outcomes and thus know that 1 + 1 = 2 (West and Young, 2002). Birds, including pigeons, chickens (even newborn chicks), and crows, are capable of calculating and estimating collections of items (e.g., Scarf et al., 2011, Ditz and Nieder, 2016, Rugani et al., 2008). Although not all experiments control for nonnumerical cues, such as the total surface area or the density of displays, many experiments have done so. For example, mosquito fish can discriminate between smaller (e.g., 3 vs. 2) and larger shoal (e.g., 8 vs. 5), even when controlling for the density of the fish and the overall space occupied by the group (Dadda et al., 2009). For larger groups (e.g., 8 vs. 4), total area and the amount of movement of the fish in both groups matter (Agrillo et al., 2008). Such experiments strongly suggest that it is numerical cues—rather than nonnumerical continuous variables—that animals are responsive to.

A recent domain of inquiry is numerical competence in insects and spiders (see Pahl et al., 2013 for a review). Although insects, such as bumblebees, have small nervous systems, they are very adept at integrating complex information, such as the relative returns of nectar by particular types of flowers, even depending on times of the day and the probability of yields (Real, 1991). Numerical information is one such source of information that insects and spiders use in their everyday ecological decisions. *Portia africana* spiders, for example, practice communal predation, sharing their prey with another resident conspecific. Juvenile *Portia africana* prefer to settle when there is one conspecific present, preferring this outcome to zero, two or three conspecifics (Nelson and Jackson, 2012). Dacke and Srinivasan (2008) designed a carefully controlled experiment where bees were trained to fly in a long tunnel where five landmarks, consisting of identical yellow strips, were placed at irregular intervals. A feeder was hidden at one of those landmarks. In the test condition, the researchers examined whether bees would look for the feeder close to the landmark where they were previously trained to find the feeder in. The bees' accuracy was very high up until landmark 3, but became more erratic at 4 and 5. Bees not only are able to discriminate numbers sequentially, but also to visually discriminate different numerosities of displays. They can match displays of two blue dots and two yellow stars, and can do so up to 3, and their performance drops at chance level at 4 (Gross et al., 2009). This is a striking similarity to other animals, suggesting bees may be subject to the same limitations of the OFS as human infants (Starkey and Cooper, 1980).

3.2 Numerical cognition plays a crucial role in animal adaptive decision making

While early authors writing on numerical competence in animals tended to dismiss it as a last resort, to which animals only turn if there is no other information available (Davis and Pérusse, 1988), the current consensus is that animals use their number sense in adaptive decision making. The best-studied ecological situation in which animals rely on numerosities is food choice:

given that they need to travel to a source of food and use up time and energy doing so, it makes sense to go to the source that has the most food. Research indicates that animals tend to "go for more," selecting maze-arms, feeders, and other experimental setups that have the largest number of food items. For example, free-ranging adult salamanders placed in a T-shaped enclosure that could choose between the ends containing either one or two live flies, or two or three live flies, chose the arm of the enclosure with the most flies (like other amphibians, salamanders can only visually see small stimuli if they move). However, they showed no preference if the choice was between 3 and 4 or 5 and 6, again revealing limits to the OFS (Uller et al., 2003).

Petroica australis, a food-caching songbird, shows sophisticated reliance on numerical information when storing, retrieving, and pilfering caches of food (mealworms). The birds could watch food being put in a pair of artificial cache sites, and could choose one of them. They were successful in finding the cache with the most mealworms (experimenters controlled for duration and other nonnumerical confounds) in caches up to twelve items. The experimenters also did a violation-of-expectation experiment, where birds watched a number of mealworms being stored, but only a subset was findable, and they examined whether these birds would take a longer time searching for the remaining worm(s). This study revealed that birds looked longer in 2 versus 1 in 3 versus 2, but not in 8 versus 4 conditions, perhaps because they were subject to the limitations of OFS which is especially operative for keeping numerosities in working memory (Hunt et al., 2008).

Animals also use numerical information for selecting their territory (Nelson and Jackson, 2012), and for choosing whether or not to attack a rivaling group, based on a comparison of that group's size and the own group's size (e.g., McComb et al., 1994 for a study with wild lionesses). Shoaling fish choose to aggregate with shoals based on their perceived size, for example, guppies prefer a shoal of 8 over a shoal of 4 to aggregate with (Bisazza et al., 2010).

3.3 Similarities in computational characteristics and limitations

As we have seen, the OFS and the ANS are the dominant ways to explain human numerical cognition. Both have specific limitations and characteristics. The OFS is accurate for collections of items up to 3 or 4. It allows animals to make comparisons and calculations across modalities. For example, rhesus monkeys (Jordan et al., 2005) and human infants (Jordan and Brannon, 2006) can match the number of voices they hear to the correct number of speaking heads they see on a monitor. It also supports addition and subtraction. Infants, as well as domestic dogs, show surprise at unexpected additions and subtractions, such as 1 + 1 = 1 or 2 − 1 = 2 (Wynn, 1992, West and Young, 2002). Above 3 or 4, the OFS is not able to make calculations or comparisons anymore. For example, chicks can discriminate between displays of 1 and 2, and between 2 and 3 items, but not between 3 and 4, or 4 versus 5, or 4 versus 6 (Rugani et al., 2008), although chicks can tell the difference between larger numbers when the ratio difference between them is large enough, for example, 2 versus 8 and 8 versus 32 (Rugani et al., 2015). We saw above that this limitation was also observed in bees, salamanders, and human infants. Human numerate adults can, of course, distinguish between collections of 3 and 4, or 4 and 6. Yet, even adults are subject to the limitations of the OFS: they are much more accurate in enumerating small collections of items (up to 3) than larger collections, with a steep decline in precision after 3 (Revkin et al., 2008). The explanation for this limitation of the OFS is that there are inherent limitations to working memory. The OFS works by putting mental representations of discrete objects (e.g., two bananas, one sound and one dot) in a placeholder format as slots that are kept in working memory (Feigenson and Carey, 2005).

The ANS, unlike the OFS, does not have a strict limit on how much it can represent, although experimental setups typically stay under 100. This system handles the approximate representation of numbers, and, like the OFS, it can work across modalities, and it supports addition and subtraction (Barth et al., 2005). Next to these features, its outputs show the Weber-Fechner signature: the discriminability of two magnitudes (numerosities) is determined by their ratio. As a result, numerical judgment improves with increasing distance (e.g., it is easier to discriminate 2 from 8 than 7 from 8, not only if this is presented as collections of dots but even in symbolic format, see Moyer and Landauer, 1967 for the first classic study to show the distance effect in symbolic format).

Comparative research indicates that rhesus monkeys' performance on approximate arithmetical tasks is similar to that of college students. Students and rhesus monkeys were required to mentally add a number of dots and select a display that showed the sum (e.g., for displays of 1 and 7 dots, the display containing 8 dots had to be selected). Next to the display showing the correct sum (e.g., 1 + 7 = 8) there was a distractor display that contained an incorrect number of dots (e.g., 1 + 7 = 5), which had a cumulative surface area close to the correct solution. Although adults were more correct (94 percent correct answers, compared to only 74 percent for the monkeys), their response patterns were very similar, showing similar sensitivity to the ratio between the numerical values of the sum and choice stimuli, in line with the Weber-Fechner law (Cantlon and Brannon, 2007). In a direct comparative study, pigeons performed on a par with primates in numerical tasks such as ordering cards with different numbers of items in ascending order, showing very similar distance effects, that is, better performance if numerosities lie further apart (Scarf et al., 2011).

There have been a few systematic studies that have examined whether the ANS in nonhuman animals other than primates obey the Weber-Fechner law. Gómez-Laplaza and Gerlai (2011) showed that angelfish (*Pterophyllum scalare*) can choose the larger of two shoals, and that their number discrimination is sensitive to the ratio difference between the two groups: for example, they prefer to aggregate with the larger shoal if the differences are 4:1 (e.g., 12 vs. 3), 3:1 (9 vs. 3), and 2:1 (8 vs. 4), but not at smaller ratio differences, for example, 1.5:1 (9 vs. 6 and 6 vs. 4). Carrion crows (*Corvus corone*) can discriminate numbers up to 30 (in displays that controlled for total surface area) in line with the Weber-Fechner law (Ditz and Nieder, 2016). While more research would need to be carried out to see how far this generalizes, the research so far supports similar cognitive mechanisms of OFS and ANS underlying animal numerical competence in a wide range of species.

3.4 Neural correlates of numerical cognition show evidence of convergent evolution

In human brains, several areas of the neocortex are associated with numerical cognition, in particular the bilateral intraparietal sulci. This area is active when adults engage in calculation with Arabic digits and dots (e.g., Dehaene et al., 1999) or even when participants are merely passively looking at or listening to Arabic digits or number words (Eger et al., 2003). The intraparietal sulci are also active in four-year-olds and in adults when presented with visual displays of collections of items that differ in number (Cantlon et al., 2006). Homologous areas in the primate parietal cortex and prefrontal cortex are responsive to numerosity. Recordings of single neuron responses in the brains of monkeys show that there are number-sensitive neurons in the lateral prefrontal cortex and the intraparietal sulcus of the posterior parietal cortex. These number-sensitive neurons selectively respond to a specific number of items in visual dot displays, including zero. Their response does not vary with other spatial features, such as the size of dots, but seems to be number-specific. While they preferentially fire at a given number of dots (say, 3), they will also respond, albeit less frequently, to other numerosities (say, 2 or 4), with response patterns following a Gaussian curve around the preferred numerosity (Tudusciuc and Nieder, 2007).

Bird numerical cognition is situated in the endbrain, more specifically in the nidopallium caudolaterale. The neurons of crows in this part of the brain fired selectively for different numerosities, just like they did in rhesus monkeys, for example, a neuron selectively tuned to 4 items also responded, but to a lesser extent, to 3 and 5 items (Ditz and Nieder, 2015). Primates and birds have markedly different brain structures. Their last common ancestor lived about 300 million years ago, at a time when the six-layered neocortex (which hosts, among others, the neurons responsible for numerical cognition) had not evolved yet in mammals. Thus, the similarities between crows and rhesus monkeys in neural representation of number show a striking convergent evolution.

The similarities between insect and mammalian (including human) numerical cognition cannot be due to homologous neural structures either. The European honeybee only weighs 0.1 g, and its brain only weighs 0.001 g, with a total size of 1 cubic mm^3, and about 1 million neurons. Compared to the human brain with its 100 billion neurons, it has only 1/100,000th of the number of human neurons. The main functions relating to memory and adaptive decision making are situated in the mushroom bodies and the central complex, so this is likely also where numerical cognition takes place. Unfortunately, it is not possible at present to find the neural correlates for numerical cognition in such a small brain (see, however, Greco et al., 2012 for recent advances in scanning brains of live bees). Given these similarities in processing, in spite of very different neural implementation, insect numerical cognition presents another case of convergent evolution.

4 The metaphysical significance of INC

The behavioral and neural similarities in the numerical cognition of a wide diversity of species and clades are a remarkable phenomenon, which I termed INC. It cannot be explained by homology (similarities due to a shared ancestral trait) given how divergent insect, avian and mammalian brains are. If homology cannot explain INC, what alternative do we have? Homology is often contrasted to homoplasy (similarity due to independent evolution), but homoplasy is a portmanteau term for several distinct evolutionary patterns (Hall, 2013). One of these is convergent evolution, when similar features evolve independently in different species as a result of similar evolutionary pressures. For example, insects, birds, and bats developed wings that help them to escape predators or pursue prey. INC is a good candidate for convergent evolution: a trait that emerged in diverse clades as a result of similar evolutionary pressures. An alternative explanation for INC is homology, when traits that evolve through convergent evolution share a similar genetic regulatory apparatus. Examples include the Pax6 gene, which helps regulate vision in mollusks, vertebrates, and insects, and the FoxP2 gene, which regulates human language development and song production in songbirds (Scharff and Petri, 2011). However, even in these cases of deep homology there is considerable independent evolution to accommodate anatomical differences (e.g., the eye structure of insects versus mammals). Moreover, if the structures in question were not adaptive, it is unlikely that these deep homologies would have occurred. For example, the Pax6 gene regulates the prenatal development of eyes, such as the iris, and its function can be explained by the fact that seeing is adaptive. Thus, even if a deep homology underlies numerical competence in these widely divergent clades, the similarities between them remain striking.

Some of the convergence in numerical cognition across clades likely has to do with constraints in computation and memory, including the Weber-Fechner signature and the limitations of the OFS. Nothing of mathematical interest happens when natural numbers become larger than 3, it is just a limitation of working memory. Why would animals be better at discriminating smaller numerosities, and why would the ratio difference be more relevant than the absolute difference? The difference between small numbers is often more ecologically significant than that between large numbers, for example, to a hungry foraging monkey, it is more relevant to see the difference between a patch with 1 fruit versus 2 fruits than that it is to be able to distinguish between 11 and 12 fruits. This ecological function of numerical cognition leads me to posit the following claim: INC presents substantial evidence for mathematical realism. It indicates that animals are tracking something in the environment (numerosities), and realism is the best explanation for numerosities.

In an earlier paper (De Cruz, 2016), I outlined an indispensability argument for mathematical realism from numerical cognition. I proposed that the best explanation for numerosities involves numbers—animals make representations of magnitude in the way they do because they are tracking structural (or other realist) properties of numbers. This fits in an ongoing discussion on whether physical phenomena have genuine mathematical explanations. Baker (2005, 2015) has argued that this is the case, citing such cases as the primeness of the life cycle of insects that are members of the genus *Magicicada* and structural properties of honeycombs. In the case of *Magicicada*, their life cycles are either 13 or 17 years. These consist of a long phase they spend as larvae underground and a brief adult phase spent above ground when they reproduce. The primeness of their life cycles makes it less likely that the life cycles of predatory species would intersect with them, thus increasing their reproductive success. Primeness is a mathematical property that plays a relevant role in the biological explanation for why their life cycles have these durations. Such examples are used to bolster the case for platonism about mathematical objects.

I will not reiterate these arguments here, but instead will consider Clarke-Doane's (2014, manuscript) more recent challenge. Now, the mathematical beliefs Clarke-Doane targets are those of professional mathematicians, such as the axioms of set theory, rather than more elementary beliefs such as that 7 is prime, or that 2 + 2 = 4. Indeed, he is happy to concede that the latter would be safe, just like the belief that burning babies for fun is wrong is true for any moral non-error-theorist (Clarke-Doane, 2014). However, the evolutionary challenge against mathematical realism targets those more basic beliefs too, just like evolutionary debunking arguments against moral realism challenge fundamental moral beliefs such as that pain is bad.

With INC we have a clear disanalogy between mathematics and morality: the proto-moral beliefs of different species are divergent, whereas numerical cognition is invariant across species. This makes numerical cognition less susceptible to the contingency challenge that has been proposed against moral realism. In the case of moral realism, one can see how our beliefs would easily have been different if our evolutionary history had gone a different way. Our moral beliefs could have easily been false (assuming a nonnaturalist form of moral realism[1]), but our mathematical beliefs could not have been. This is because evolution has shaped our minds (as well as those of bees, crows, rhesus monkeys, chicks, angelfish, etc.) to track numerical information. Similarities in numerical cognition across a wide range of unrelated species require some explanation, and mathematical realism can provide this explanation straightforwardly, namely what animals are tracking are mathematical truths/structures. I am not arguing that antirealists cannot explain INC. Nevertheless, the antirealist would need to explain why unrelated animals such as salamanders, bees, crows, angelfish, and rhesus monkeys (and of course humans), would be able to track discrete quantities in their environment, would be able to do so across modalities, and would use this information to inform their adaptive choices. INC thus shifts the burden of proof in the direction of the antirealist.

One can, of course, resort to highly contrived scenarios where animals have adaptive responses, such as choosing the most numerous shoal or cache of mealworms, without tracking mathematical truths. Plantinga's (1993) evolutionary argument against naturalism famously argued that animals can have the right adaptive behaviors without truth-tracking beliefs, for example, a hominin who runs away from a tiger (adaptive response), but does so because he believes the tiger is cute and he wants to pet it, but he also believes that the best way to pet it is to run away from it (maladaptive belief). While such scenarios are metaphysically possible (and some have outlined them for the case of numerical beliefs, for example, Clarke-Doane, 2012), they are not very plausible. An error-theorist would have to come up with a scenario for each case of evolved numerical cognition (which, to the best of our knowledge, has occurred independently at least in insects, birds, mammals, and fish) where somehow wrong or irrelevant mathematical beliefs would lead to the right adaptive responses. At present, there is no satisfying positive case for mathematical antirealism that accounts for INC without resorting to arcane scenarios.

5 Which form of realism does the animal cognition literature support?

The Benacerraf-Field challenge to mathematical realism asks realists to explain the reliability of mathematical beliefs. This is not as demanding as outlining a causal account (which would be impossible under some forms of realism in any case), but requires us to show that our evolved mathematical beliefs are safe from error. I have argued in previous work (De Cruz, 2016) that Shapiro's (1997) *ante rem* structuralism is a possible candidate in the light of evolution. One reason to look more closely into realist structuralist accounts is that authors in this field, such as Shapiro (1997), have made substantial efforts to explain how their account would work in a naturalistic framework. Moreover, *ante rem* structuralism provides a straightforward account of reference and semantics, and can provide an account of mathematical structure irrespective of the agent cognizing it, which makes the approach suitable for our explanation of numerical cognition across species. To summarize, *ante rem* structuralism holds that non-applied mathematics is concerned with structures that are conceived of as abstract entities (platonic universals), that is, structures that exist independently and prior to any instantiations of them. *Ante rem* structuralists do not specify the precise nature of these entities, but rather focus on the role they play. Numbers are positions in a certain structure, and can be discerned in the environment as patterns. The bootstrapping account (Carey, 2009) can offer a glimpse of how we can have mathematical beliefs that are safe, and that track mathematical structures. According to this account, young children learn to recognize the 1, 2, and 3 pattern, thanks to their OFS, which allows for exact discrimination of numerosities up to 3. Since the OFS is very precise, learning the 1, 2, and 3 pattern is a reliable process (at least in neurotypical children who do not suffer from dyscalculia). The children learn the remaining natural numbers through a process of induction. This part of the learning process is stable thanks to the abundant cultural scaffolding (e.g., counting songs) and feedback (e.g., parents correcting their child, or helping their child to count a given collection of items) children receive (see also De Cruz, 2018). In this way, an *ante rem* structuralist can explain the reliability of our natural number concepts to track mathematical truths.

Another plausible realist (non-platonist, in this case) account that is compatible with INS, is Millean empiricism. Kitcher (1984) revived this position, arguing that mathematical epistemology should seek inspiration from how children learn arithmetic. It thus fits well in a naturalistic account of mathematical cognition. A closely related view is Aristotelian realism, recently defended by Franklin (2014). Mill (1843, 165) proposed that numbers are properties of physical aggregates:

> When we call a collection of objects two, three, or four, they are not two, three, or four in the abstract; they are two, three, or four things of some particular kind; pebbles, horses, inches, pounds' weight. What the name of number connotes is, the manner in which single objects of the given kind must be put together, in order to produce that particular aggregate.

In this view, numerical cognition detects high-level, general properties of aggregates, for example, an angelfish that chooses a shoal of 7 fish over 3 fish is detecting the high-level general property of aggregates that $7 > 3$. According to Mill, we do not need to invoke the existence of 3 and 7, separate from their concrete instantiations in the physical world. Millean empiricism does not presuppose platonist ontology, but it is nevertheless a realist ontology (see Balaguer, 1998, chapter 5), because it regards the laws of arithmetic as highly as general laws of nature.

A common objection to Millean empiricism is that aggregates do not have determinate number properties. For example, a group of lions can be divided into many different parts, for instance, it is composed of 7 lions, 28 legs, etc. Kessler (1980) responds to this problem by arguing that in Mill's account, numbers are not properties of aggregates, but relations that hold between aggregates (e.g., the pride) and properties of those aggregates (e.g., individual lions). Infants and animals are successful at finding the relevant properties of aggregates in numerical tasks, for instance, they can compare the number of speakers they see with the number of voices they hear (Jordan and Brannon, 2006). When they are presented with a collection of objects (e.g., an array of dots) infants seem to be less able to detect a decrease or increase in the individual object's size, than they are to detect a change in numerosity. They need as much as a fourfold change in size to notice it, as revealed by a longer looking time. This suggests that once infants attend to numerosity, they disregard the physical particulars of the items that constitute them (Cordes and Brannon, 2011). In line with Millean empiricism, they can make high-level generalizations about numerosities that go beyond the physical properties of aggregates. Given that the world at our scale mostly consists of separable objects, there may have been an evolutionary advantage of making high-level generalizations about numerosities, along the lines of separable objects as we and other animals encounter them in daily life (see also Dehaene, 2011: 231).

Ante rem structuralism and Millean empiricism are two realist ontologies that are compatible with the evolved features of numerical cognition. Both meet the Benacerraf-Field challenge of explaining the reliability of numerical representations. For the structuralist account, direct interaction with structures is not required to know numbers, and for Millean empiricism, numerosities form a high-level generalization of the properties of discrete middle-sized objects.

6 Conclusion

In this chapter, I have argued that mathematics and morality are disanalogous in an important respect. Mathematical beliefs seem to be less contingent upon our peculiar evolutionary history than moral beliefs are. I have presented evidence for invariantism in numerical cognition: numerical cognition occurs across many animal clades, including insects, fish, amphibians, birds, and mammals, and is, according to the dominant theories on numerical cognition, subserved by two systems: the ANS, which deals with larger collections of items through approximation, and the OFS, an exact system for small numerosities up to 3 or 4. Numerical information plays a crucial role in animal decision making. Animals across widely different clades show similar cognitive limitations and strategies in dealing with numbers, including an ability to deal with numbers across modalities. Neural evidence suggests multiple instances of convergent evolution. If animal minds have hit upon these solutions so many times independently, this would be a formidable coincidence which antirealists would need to explain. Of course, INC also requires an explanation under the assumption of mathematical realism. In particular, the Benacerraf-Field challenge asks the mathematical realist to explain the reliability of mathematical beliefs. If this were in principle impossible to achieve, this would undermine our mathematical beliefs, according to Field (1989). The Benacerraf-Field challenge can be cashed out in terms of safety: the realist needs to show that we could not easily have had false mathematical beliefs.

I showed that ante rem structuralism and Millean empiricism provide a solution to the Field-Benacerraf challenge: they can explain the reliability of animal numerical beliefs, and thus by extension of human mathematical beliefs that are based upon them, such as the belief that 11 follows 10, or that 7 is prime. My argument does not provide an evolutionary justification of more formal mathematical beliefs, such as those involved in set theory.

Acknowledgments

Many thanks to Johan De Smedt, Justin Clarke-Doane, Brendan Larvor, Jan Verpooten, Anne Jacobson, two anonymous reviewers, and the editors of this volume for their comments to an earlier version of this chapter.

Note

1 Some naturalistic forms of moral realism are less susceptible to the contingency challenge, in particular the neo-Aristotelian approach to morality (as e.g., outlined by Foot, 2001). According to neo-Aristotelians, what counts as a good human life and human flourishing is the truth-maker of moral claims. Humans have, as evolved creatures, certain limitations on the conditions that will make them thrive and flourish. The role of the ethicist is to find out how to fulfill these conditions. Fitzpatrick (2000) has challenged the neo-Aristotelian account by pointing out that not all evolved features lead to flourishing; for example, male elephant seals fight to gain control of large harems, which makes evolutionary sense but does not seem to contribute to their well-being. However, as Lott (2012) has countered, the neo-Aristotelian approach does not look at animals from the outside, but instead from the inside of life forms. In that respect, it would seem that it is "good" for bee queens to kill their fertile daughters, to harken back to Darwin's example.

References

Agrillo, C., Dadda, M., Serena, G. and Bisazza, A. (2008). Do fish count? Spontaneous discrimination of quantity in female mosquitofish. *Animal Cognition*, 11(3):495–503.

Baker, A. (2005). Are there genuine mathematical explanations of physical phenomena? *Mind*, 114:223–38.

Baker, A. (2015). Mathematical explanation in biology. In P.-A. Braillard and C. Malaterre (Eds.), *Explanation in Biology*, 229–47. Dordrecht, The Netherlands: Springer.

Balaguer, M. (1998). *Platonism and Anti-Platonism in Mathematics*. New York: Oxford University Press.

Barth, H., La Mont, K., Lipton, J. and Spelke, E. S. (2005). Abstract number and arithmetic in preschool children. *Proceedings of the National Academy of Sciences of the United States of America*, 102(39):14116–21.

Benacerraf, P. (1973). Mathematical truth. *Journal of Philosophy*, 70:661–80.

Bisazza, A., Piffer, L., Serena, G. and Agrillo, C. (2010). Ontogeny of numerical abilities in fish. *PLoS One*, 5: e15516.

Bogardus, T. (2014). Knowledge under threat. *Philosophy and Phenomenological Research*, 88(2):289–313.

Brannon, E. M. and Terrace, H. S. (1998). Ordering of the numerosities 1 to 9 by monkeys. *Science*, 282(5389):746–49.

Cantlon, J. F. and Brannon, E. M. (2007). Basic math in monkeys and college students. *PLoS Biology*, 5(12):e328.

Cantlon, J. F., Brannon, E. M., Carter, E. J., & Pelphrey, K. A. (2006). Functional imaging of numerical processing in adults and 4-y-old children. *PLoS Biology*, 4(5): e125.

Carey, S. (2009). *The Origin of Concepts*. Oxford: Oxford University Press.

Clarke-Doane, J. (2012). Morality and mathematics: The evolutionary challenge. *Ethics*, 122:313–40.

Clarke-Doane, J. (2014). Moral epistemology: The mathematics analogy. *Noûs*, 48:238–55.

Clarke-Doane, J. (2016). What is the Benacerraf problem? In F. Pataut (Ed.), *Truth, Objects, Infinity, Logic, Epistemology, and the Unity of Science*, 17–43. Dordrecht, The Netherlands: Springer.

Clarke-Doane, J. (manuscript). Benacerraf, Pluralism, and Normativity.

Cordes, S. and Brannon, E. M. (2011). Attending to one of many: When infants are surprisingly poor at discriminating an item's size. *Frontiers in Psychology*, 2:1–8.

Dacke, M. and Srinivasan, M. V. (2008). Evidence for counting in insects. *Animal Cognition*, 11:683–89.

Dadda, M., Piffer, L., Agrillo, C. and Bisazza, A. (2009). Spontaneous number representation in mosquitofish. *Cognition*, 112:343–48.

Darwin, C. (1871). *The Descent of Man, and Selection in Relation to Sex*, Volume 1. London: John Murray.

Davis, H. and Pérusse, R. (1988). Numerical competence in animals. Definitional issues, current evidence, and a new research agenda. *Behavioral and Brain Sciences*, 11:561–615.

De Cruz, H. (2016). Numerical cognition and mathematical realism. *Philosophers' Imprint*, 16:1–13.

De Cruz, H. (2018). Testimony and children's acquisition of number concepts. In S. Bangu (Ed.), *Naturalizing Logico-Mathematical Knowledge*, 164–77. London and New York, NY: Routledge.

Dehaene, S. (2011). *The Number Sense. How the Mind Creates Mathematics*, Revised and expanded edition. New York, NY: Oxford University Press.

Dehaene, S., Spelke, E. S., Pinel, P., Stanescu, R. and Tsivkin, S. (1999). Sources of mathematical thinking: Behavioral and brain-imaging evidence. *Science*, 284:970–74.

Ditz, H. M. and Nieder, A. (2015). Neurons selective to the number of visual items in the corvid songbird endbrain. *Proceedings of the National Academy of Sciences*, 112:7827–32.

Ditz, H. M. and Nieder, A. (2016). Numerosity representations in crows obey the Weber–Fechner law. *Proceedings of the Royal Society B*, 283:20160083.

Eger, E., Sterzer, P., Russ, M. O., Giraud, A.-L. and Kleinschmidt, A. (2003). A supramodal number representation in human intraparietal cortex. *Neuron*, 37:1–20.

Feigenson, L. and Carey, S. (2005). On the limits of infants' quantification of small object arrays. *Cognition*, 97:295–313.

Feigenson, L., Dehaene, S. and Spelke, E. S. (2004). Core systems of number. *Trends in Cognitive Sciences*, 8:307–14.

Field, H. (1989). *Realism, Mathematics, and Modality*. Oxford: Blackwell.

Fitzpatrick, W. J. (2000). *Teleology and the Norms of Nature*. New York, NY: Garland Publishing.

Foot, P. (2001). *Natural Goodness*. Oxford: Oxford University Press.

Franklin, J. (2014). *An Aristotelian Realist Philosophy of Mathematics. Mathematics as the Science of Quantity and Structure*. Houndmills: Palgrave Macmillan.

Gebuis, T., Kadosh, R. C., & Gevers, W. (2016). Sensory-integration system rather than approximate number system underlies numerosity processing: A critical review. *Acta Psychologica*, 171:17–35.

Gómez-Laplaza, L. M. and Gerlai, R. (2011). Can angelfish (*Pterophyllum scalare*) count? Discrimination between different shoal sizes follows Weber's law. *Animal Cognition*, 14(1):1–9.

Greco, M. K., Tong, J., Soleimani, M., Bell, D. and Schäfer, M. O. (2012). Imaging live bee brains using minimally-invasive diagnostic radioentomology. *Journal of Insect Science*, 12:89.

Gross, H. J., Pahl, M., Si, A., Zhu, H., Tautz, J. and Zhang, S. (2009). Number-based visual generalisation in the honeybee. *PLoS One*, 4:e4263.

Hall, B. K. (2013). Homology, homoplasy, novelty, and behavior. *Developmental Psychobiology*, 55(1):4–12.

Henrich, J., McElreath, R., Barr, A., Ensimger, J., Barrett, C., Bolyanatz, A., Cardenas, J. C., Gurven, M., Gwako, E., Henrich, N., Lesorogol, C., Marlowe, F., Tracer, D. and Ziker, J. (2006). Costly punishment across human societies. *Science*, 312 (5781):1767–70.

Hunt, S., Low, J. and Burns, K. C. (2008). Adaptive numerical competency in a food hoarding songbird. *Proceedings of the Royal Society of London B: Biological Sciences*, 275(1649):2373–79.

Jordan, K. E. and Brannon, E. M. (2006). The multisensory representation of number in infancy. *Proceedings of the National Academy of Sciences of the United States of America*, 103:3486–89.

Jordan, K. E., Brannon, E. M., Logothetis, N. K. and Ghazanfar, A. A. (2005). Monkeys match the number of voices they hear to the number of faces they see. *Current Biology*, 15:1–5.

Joyce, R. (2006). Metaethics and the empirical sciences. *Philosophical Explorations*, 9:133–48.

Kessler, G. (1980). Frege, Mill, and the foundations of arithmetic. *Journal of Philosophy*, 77:65–79.

Kitcher, P. (1984). *The Nature of Mathematical Knowledge*. New York, NY & Oxford: Oxford University Press.

Lillehammer, H. (2010). Methods of ethics and the descent of man: Darwin and Sidgwick on ethics and evolution. *Biology & Philosophy*, 25(3):361–78.

Lott, M. (2012). Have elephant seals refuted Aristotle? Nature, function, and moral goodness. *Journal of Moral Philosophy*, 9(3):353–75.

Lourenco, S. F., Bonny, J. W., Fernandez, E. P., & Rao, S. (2012). Nonsymbolic number and cumulative area representations contribute shared and unique variance to symbolic math competence. *Proceedings of the National Academy of Sciences*, 109(46): 18737–42.

McComb, K., Packer, C. and Pusey, A. E. (1994). Roaring and numerical assessment in contests between groups of female lions. *Panthera leo. Animal Behaviour*, 47:379–87.

Mill, J. S. (1843). *A System of Logic, Ratiocinative and Inductive, Being a Connected View of the Principles of Evidence, and the Methods of Scientific Investigation*, Volume 2. London: John W. Parker.

Moyer, R. S. and Landauer, T. K. (1967). Time required for judgements of numerical inequality. *Nature*, 215:1519–20.

Nelson, X. J. and Jackson, R. R. (2012). The role of numerical competence in a specialized predatory strategy of an araneophagic spider. *Animal Cognition*, 15(4):699–710.

Pahl, M., Si, A. and Zhang, S. (2013). Numerical cognition in bees and other insects. *Frontiers in Psychology*, 1–9.

Pica, P., Lemer, C., Izard, V. and Dehaene, S. (2004). Exact and approximate arithmetic in an Amazonian indigene group. *Science*, 306:499–503.

Plantinga, A. (1993). *Warrant and Proper Function*. Oxford: Oxford University Press.

Real, L. A. (1991). Animal choice behavior and the evolution of cognitive architecture. *Science*, 253:980–86.

Revkin, S. K., Piazza, M., Izard, V., Cohen, L. and Dehaene, S. (2008). Does subitizing reflect numerical estimation? *Psychological Science*, 19:607–14.

Rips, L. J., Asmuth, J. and Bloomfield, A. (2006). Giving the boot to the bootstrap: How not to learn the natural numbers. *Cognition*, 101:B51–60.

Rips, L. J., Bloomfield, A. and Asmuth, J. (2008). From numerical concepts to concepts of number. *Behavioral and Brain Sciences*, 31:623–42.

Rowland, R. 2016. Rescuing companions in guilt arguments. *The Philosophical Quarterly*, 66:161–71.

Rugani, R., Regolin, L. and Vallortigara, G. (2008). Discrimination of small numerosities in young chicks. *Journal of Experimental Psychology: Animal Behavior Processes*, 34(3):388–99.

Rugani, R., Vallortigara, G., Priftis, K. and Regolin, L. (2015). Number-space mapping in the newborn chick resembles humans' mental number line. *Science*, 347(6221):534–36.

Scarf, D., Hayne, H. and Colombo, M. (2011). Pigeons on par with primates in numerical competence. *Science*, 334:1664.

Scharff, C. and Petri, J. (2011). Evo-devo, deep homology and FoxP2: Implications for the evolution of speech and language. *Philosophical Transactions of the Royal Society of London B: Biological Sciences*, 366(1574):2124–40.

Schneider, M., Beeres, K., Coban, L., Merz, S., Susan Schmidt, S., Stricker, J. and De Smedt, B. (2017). Associations of non-symbolic and symbolic numerical magnitude processing with mathematical competence: A meta-analysis. *Developmental Science*, 20(3):e12372.

Shapiro, S. (1997). *Philosophy of Mathematics: Structure and Ontology*. Oxford: Oxford University Press.

Silk, J. B. and House, B. R. (2011). Evolutionary foundations of human prosocial sentiments. *Proceedings of the National Academy of Sciences*, 108(Supplement 2):10910–17.

Spelke, E. S. and Kinzler, K. D. (2007). Core knowledge. *Developmental Science*, 10:89–96.

Starkey, P. and Cooper, R. G. (1980). Perception of numbers by human infants. *Science*, 210:1033–35.

Street, S. (2006). A Darwinian dilemma for realist theories of value. *Philosophical Studies*, 127:109–66.

Street, S. (2008). Reply to Copp: Naturalism, normativity, and the varieties of realism worth worrying about. *Philosophical Issues*, 18:207–28.

Tudusciuc, O. and Nieder, A. (2007). Neuronal population coding of continuous and discrete quantity in the primate posterior parietal cortex. *Proceedings of the National Academy of Sciences USA*, 104:14513–18.

Uller, C., Jaeger, R., Guidry, G. and Martin, C. (2003). Salamanders (*Plethodon cinereus*) go for more: Rudiments of number in an amphibian. *Animal Cognition*, 6:105–12.

Vonk, J. and Beran, M. J. (2012). Bears "count" too: Quantity estimation and comparison in black bears, *Ursus americanus*. *Animal Behaviour*, 84(1):231–38.

West, R. E. and Young, R. J. (2002). Do domestic dogs show any evidence of being able to count? *Animal Cognition*, 5(3):183–86.

West, S. A. and Gardner, A. (2010). Altruism, spite, and greenbeards. *Science*, 327:1341–44.

Wynn, K. (1992). Addition and subtraction by human infants. *Nature*, 358:749–50.

4

The Beauty (?) of Mathematical Proofs

Catarina Dutilh Novaes

Mathematicians often use aesthetic vocabulary to describe mathematical proofs: they can be beautiful, elegant, ugly, etc. In recent years, philosophers of mathematics have been asking themselves what these descriptions in fact mean: should we take them literally, as tracking truly aesthetic properties of mathematical proofs, or are these terms being used as proxy for non-aesthetic properties, in particular epistemic properties? Starting from the idea that one of the main functions of mathematical proofs is to explain and persuade an interlocutor, I develop an account of the beauty (or ugliness) of mathematical proofs that seems to allow for a reconciliation of these apparently opposed accounts of aesthetic judgments in mathematics. I do so by discussing the role of *affective responses and emotions* in the practice of mathematical proofs, thus arguing that the aesthetic and the epistemic are intrinsically related (while not entirely coinciding) in mathematical proofs.

1 Introduction

It is well-known that mathematicians often employ aesthetic vocabulary to describe mathematical entities, mathematical proofs in particular.[1] Poincaré claimed that mathematical beauty is "a real aesthetic feeling that all true mathematicians recognize." (Poincaré, 1930: 59) In a similar vein, Hardy remarked that "beauty is the first test: there is no permanent place in the world for ugly mathematics" (Hardy, 1940: 14). More recently, corpus analysis of the laudatory texts on the occasion of mathematical prizes such as the Abel prize shows that they are filled with aesthetic terminology (Holden and Piene, 2009). Moreover, certain kinds of proofs that (still) encounter resistance among mathematicians, such as computer-assisted proofs or probabilistic proofs, are often described as "ugly" (Montaño, 2012). Indeed, mathematicians seem to use aesthetic vocabulary systematically to indicate their *predilection* for some proofs over others.

What exactly is going on? Even if we keep in mind that, in colloquial language, it is quite common to use aesthetic terminology in a rather loose sense ("she has a beautiful mind"; "things got quite ugly at that point"), the robustness of uses of this terminology among mathematicians seems to call for a philosophical explanation. What are these judgments tracking? Are they tracking truly *aesthetic* properties of mathematical proofs? Or are these aesthetic terms being used as proxy for some other, non-aesthetic property or properties? Is it really the case that "all true mathematicians" recognize mathematical beauty when they see it? Do they converge in their attributions of beauty (or ugliness) to mathematical proofs? And even assuming that there is a truly aesthetic dimension in these judgments, is beauty a property of the proofs themselves, or is it rather something "in the eyes of the beholder"? These are some of the explanatory challenges for the philosopher of mathematics seeking to understand why mathematicians systematically employ aesthetic terminology to talk about mathematical proofs, and what this may mean for a philosophical account of mathematical proofs.

In the literature on the beauty of proofs, there seem to be two main kinds of accounts of uses of aesthetic vocabulary by mathematicians: the literal, non-reductive accounts; and the nonliteral, reductive accounts.[2] According to literal, non-reductive accounts, when mathematicians use aesthetic terms to talk about proofs, they truly mean what they say; they are talking about genuinely *aesthetic* properties (which may be in the proofs themselves or not, but are in any case genuinely aesthetic). Accounts of this kind may well attempt to explicate the idea of beauty in terms of other concepts,[3] but they attribute an irreducibly aesthetic dimension to these judgments. In contrast, according to the nonliteral, reductive accounts, when mathematicians use terms such as "beautiful," "pleasing," "ugly," etc. to describe proofs, these terms are being used as proxy for non-aesthetic features. In other words, according to reductive accounts, these aesthetic adjectives are merely a *façon de parler*; they are not tracking truly aesthetic properties at all. More precisely, reductivists typically claim that the properties being tracked are in fact purely *epistemic*, not aesthetic. The presupposition in these debates seems to be that there is a real opposition between epistemic and aesthetic dimensions of a mathematical proof.

In this chapter, I develop an account of the beauty (or ugliness) of mathematical proofs that seems to offer a possibility of reconciliation for these apparently opposed accounts of aesthetic judgments in mathematics.[4] I do so by discussing the role of *affective responses and emotions* in the practice of mathematical proofs, thus arguing that the aesthetic and the epistemic are intrinsically related (while not entirely coinciding). To this end, I rely on Hardy's list of six properties of a beautiful mathematical proof as described in Hardy (1940) (which I will refer to as "Hardy's Six"), and on the dialogical conception of proofs and deductive reasoning that I have been developing in recent years (Dutilh Novaes, 2013, 2016, 2018), where a proof's *persuasive function* plays an important role. Along the way, I address a number of methodological questions on how these issues should be addressed, and argue for the significance of a combination of empirical/experimental methods.

I proceed in the following way. I begin with a brief survey of the literature, emphasizing the literal/non-reductive versus nonliteral/reductive opposition (see "Overview of literature"). I then discuss a concrete example of a purportedly beautiful proof, Cantor's diagonal argument, relying on Hardy's Six (see "A beautiful proof"). Next, I argue that five of Hardy's Six can be understood in terms of the notion of the explanatoriness of a proof, a quintessentially epistemic property (see "Hardy's Six and explanatoriness"). We are left with one of them, unexpectedness, that does not quite seem to fit within the explanatoriness story. In "Dialogical conception: Explanatory persuasion," I present some of the details of my dialogical conception of mathematical proofs in order to motivate the centrality of persuasiveness for proofs. In "The aesthetic import of unexpectedness," I compare the role of unexpectedness in mathematical proofs to its role in poetry and music. Next (see "Unexpectedness and the role of affective responses"), I discuss some empirical work on the connections between unexpectedness, affective responses, aesthetic pleasure, and conviction/persuasion, as well as the notion of "aesthetic-epistemic feelings" introduced in Todd (2018). (To be clear, I take unexpectedness to be one among a number of properties where epistemic and aesthetic components come together. But for reasons of space, I restrict myself to unexpectedness here.) These observations lead me to the conclusion that there is no real dichotomy between aesthetic and epistemic dimensions in a proof; at the very least, there is significant overlap between them, even if they do not entirely coincide. In "Empirical predictions," I spell out some of the empirical predictions emerging from the account.

2 Overview of literature

There is a fairly significant body of literature on the presumed aesthetic dimension of mathematical proofs. For reasons of space, here I focus specifically on the question whether aesthetic vocabulary is being used to track truly aesthetic properties (literal/non-reductive accounts), or whether it is being used as proxy for some non-aesthetic properties (nonliteral/reductive accounts).

Hardy (1940) belongs to the literal, non-reductive camp, even if he explicates the beauty of proofs in terms of other concepts. According to him, a mathematical proof is beautiful if it scores high on the following dimensions: seriousness, generality, depth, unexpectedness, inevitability, and economy (Hardy's Six). Similarly, McAllister, a non-reductionist, highlights brevity and simplicity[5] as the key features of a beautiful mathematical proof:

> Mathematicians have customarily regarded a proof as beautiful if it conformed to the classical ideals of brevity and simplicity. The most important determinant of a proof's perceived beauty is thus the degree to which it lends itself to being grasped in a single act of mental apprehension. (McAllister, 2005: 22)

Another resolutely aesthetic approach to the presumed beauty (or ugliness) of a mathematical proof is developed by Montaño (2014), who draws, among others, on theories of musical aesthetics to develop a literal account of mathematical beauty (and ugliness). These approaches seek to take what mathematicians say at face value, treating their uses of aesthetic terminology as suggesting that a truly aesthetic approach is required to understand this aspect of mathematical practice. (See (Inglis and Aberdein, 2015) for further examples of authors adopting the literal, non-reductive position.)

Reductive, nonliteral approaches, in turn, posit that aesthetic terminology as used by mathematicians is in fact tracking some non-aesthetic property or properties of proofs. One exponent of this position is the mathematician Giancarlo Rota (1997), for whom the key property being tracked by these apparently aesthetic judgments is in fact a purely epistemic, non-aesthetic property: *enlightenment*.

> We acknowledge a theorem's beauty when we see how the theorem "fits" in its place,[6] how it sheds light around itself, like a *Lichtung*, a clearing in the woods. We say that a proof is beautiful when such a proof finally gives away the secret of the theorem, when it leads us to perceive the actual, not the logical inevitability of the statement that is being proved. (Rota, 1997: 182)

He then gives a quasi-sociological reason for aesthetic terms being used as proxy for the epistemic property of enlightenment:

> The term "mathematical beauty" [is a trick] that mathematicians have devised to avoid facing up to the messy phenomenon of enlightenment. . . . All talk of mathematical beauty is a copout from confronting the logic of enlightenment, a copout that is intended to keep our formal description of mathematics as close as possible to the description of a mechanism. (Rota, 1997: 182)

Rota's description of enlightenment seems to come quite close to what is commonly referred to as the property of *explanatoriness* (Mancosu and Pincock, 2012), which is thus a plausible candidate as the main epistemic property being tracked by these apparently aesthetic judgments. Indeed, it seems that aesthetic vocabulary is more often present in how mathematicians talk about proofs than explanation-related vocabulary, and this is exactly what Rota would predict: mathematicians would eschew vocabulary related to explanation and enlightenment and use aesthetic vocabulary instead to talk about these epistemic properties.[7] In a similar vein, Todd (2008) contends that the prima facie aesthetic judgments of mathematicians are in fact epistemic judgments,[8] and thus that aesthetic vocabulary is being used as proxy for epistemic assessment.[9]

But what could possibly count as evidence to adjudicate the "dispute" between the literal/non-reductive camp and the nonliteral/reductive camp? We are now confronted with a rather serious methodological challenge, namely that of determining what counts as "data" on this matter (and potentially other issues in the philosophy of mathematics). Both sides seem to have compelling arguments, but it is not clear that a top-down approach based on conceptual, philosophical argumentation alone will be sufficient.[10] However, merely anecdotal evidence and self-reported experiences of mathematicians will not suffice either.[11] First, there are well-known limits to introspective knowledge, so it is not obvious that the testimonies of mathematicians themselves will be telling the full story. Secondly, what is to rule out that some mathematicians use aesthetic terminology as proxy for epistemic properties, while others use the terminology in a literal sense instead? It is not clear that a uniform account is what we should aim for, that is if the target phenomena themselves may well be highly heterogeneous.[12]

The recent work of Inglis and Aberdein (2015, 2016, 2017) presents a novel methodological approach to these issues. In their initial study, they surveyed the proof appraisals of a group of self-selected mathematicians on a number of aspects, including aesthetic aspects. The instruction of the experiment was as follows:

> Please think of a particular proof in a paper or book which you have recently refereed or read. Keeping this specific proof in mind, please use the rating scale below to describe how accurately each word in the table below describes the proof. Describe the proof as it was written, not how it could be written if improved or adapted. . . . Please read each word carefully, and then select the option that corresponds to how well you think it describes the proof. (Inglis and Aberdein, 2015: 95)

Participants were then given a list of eighty adjectives in random order, such as: beautiful, enlightening, simple, profound, pleasant, explanatory, useful, charming, fruitful etc. The goal was to establish which (clusters of) adjectives were regularly used together to describe specific proofs (importantly, a different proof for each participant). Using a method borrowed from social psychology to study personality traits, from their data set Inglis and Aberdein established that participants' appraisals of proofs varied on four main dimensions: (in their terms) aesthetics, intricacy, utility, and precision. Interestingly, they found no significant correlation between ascriptions of simplicity and ascriptions of beauty to proofs in the participants' responses. This seems, at least prima facie, to count as evidence against McAllister's account of beauty in terms of simplicity. Moreover, "beautiful" correlated quite strongly with "enlightening," which seems to lend support to Rota's account, but it did not correlate as strongly with "explanatory." Inglis and Aberdein conclude,

> One interpretation of these data would be to suggest that epistemic and aesthetic judgements of mathematical proofs are indeed different. Epistemic judgements concentrate largely on the utility dimension, whereas aesthetic judgements concentrate on the aesthetics dimension. Nevertheless, there do appear to be adjectives which reside at the conjunction of these two dimensions. (Inglis and Aberdein, 2015: 103)

In other words, the nonliteral account of "beauty as proxy for something else" (for epistemic features in particular) is not fully vindicated by the analysis. But these results do suggest important connections between aesthetic and epistemic assessments.

How much can this approach tell us about the nature of aesthetic judgments in mathematics? A number of methodological objections can be raised to its philosophical significance, for example: How reliable is data on colloquial uses of aesthetic terminology? What about differences among different mathematical "cultures and traditions" in function of geography, subareas of mathematics, etc.?[13] It may be argued that, while this approach provides extremely valuable data, it does not exhaust the significance of the phenomena in question. Indeed, it seems that we must also pay attention to the underlying psychological experiences that lead mathematicians to make prima facie aesthetic assessments of proofs (which will be discussed under "Unexpectedness and the role of affective responses"). At any rate, the issue of the (presumed) beauty of mathematical proofs is a quintessential example of an issue in the philosophy of mathematics that screams for empirically informed approaches.[14]

3 A beautiful proof

At this point, it may prove useful to offer a concrete example of a proof that is often viewed as mesmerizing: Cantor's diagonal argument. In order to come to grips with its (presumed) beauty, let us turn to Hardy's six features of a beautiful proof, as described in Hardy (1940). Recall that Hardy's is not a reductive account—that is, he does not seek to reduce beauty to non-aesthetic properties, but to explicate the nature of mathematical beauty as such. Notice also that by relying on Hardy's Six, I am not claiming that they *exhaust* all there is to be said about the (presumed) beauty of mathematical proofs; there may well be facets of mathematical beauty that do not fall under any of these six categories. Still, Hardy's Six seem to offer a suitable, fairly concrete framework to address the issue.[15]

According to Hardy, a beautiful mathematical proof displays all or at least many of the following characteristics:

- **Serious**: it is connected to other mathematical ideas (p. 25)
- **General**: the main idea is used in proofs of different kinds (p. 25)
- **Deep**: pertaining to deeper "strata" of mathematical entities (p. 27)[16]
- **Unexpected**: the argument takes a surprising form (p. 29)
- **Inevitable**: there is no escape from the conclusion (p. 29)
- **Economical** (simple): there are no complications of detail, one line of attack (p. 29)

Now let us see how Cantor's diagonal argument fares with respect to these features. For the present purposes, it is sufficient to rely on a simplified, rather informal version of the proof, roughly as found in Bellos (2010: chapter 11). Recall that Cantor's diagonal argument is a proof of the uncountability of the real numbers, which means that there are more real numbers than there are natural numbers (and thus that there are different sizes of infinity).

The proof proceeds by contradiction. Suppose that the set of real numbers between 0 and 1 (denoted as (0, 1)) is countable.[17] We then enumerate all its decimal expansions[18] by fixing a bijection $f : N \to (0, 1)$ between the natural numbers and these decimal expansions (in no particular order):

1. 0,148930 . . .
2. 0,483886 . . .
3. 0,873205 . . .
4. 0,759783 . . .

. . .

By diagonalization, we obtain the number D = 0,1837 . . . , where the first number in the decimal sequence is the first number of the first item on our list, the second is the second number of the second item on our list, etc. (indicated in bold face). We then perform a uniform operation on D, for example, adding 1 to each of the numbers in the decimal series. We obtain D* = 0,2948 . . . Now, D* cannot be on our list, given that it differs from each number on the list for at least one decimal position. We thus obtain a contradiction, since we had assumed that *all* decimal expansions between 0 and 1 were on our list. From the contradiction, we conclude that the initial assumption is false, and thus that the set of real numbers between 0 and 1 is not countable.

How does Cantor's argument fare on Hardy's criteria? Let us discuss them one by one:

- **Serious**: the proof is serious since it deals with key mathematical ideas such as numbers, sets, and infinity.
- **General**: the proof is general because the technique of diagonalization, first introduced by Cantor for this proof, has since been used in a number of very important proofs (e.g., Gödel's incompleteness theorems).
- **Deep**: the proof deals with foundational entities such as the real numbers and infinity, and so is arguably deep.
- **Unexpected**: the argument does take a surprising form, when the newly created number D* turns out not to be on the list. There is, we might say, a "plot twist" at this point.
- **Inevitable**: this is the one feature on Hardy's list that cannot straightforwardly be attributed to Cantor's diagonal argument. Judging from

my teaching experience, a common reaction to the proof is the feeling that there is something "fishy" going on; quite some people do not feel the "pull" of the conclusion from the argument.[19]
- **Economical** (simple): the proof is exemplary when it comes to simplicity and brevity; a powerful and foundational result emerging from one simple idea.

In conclusion, Cantor's diagonal argument seems to score high on at least five of Hardy's Six, and so there seems to be convergence between the instinctive response of attributing beauty to this proof[20] and the analysis in terms of Hardy's criteria.[21] As I will argue later on, this proof is particularly interesting for our purposes because it tends to provoke an "aha-experience" when number D^* is introduced and turns out not to be on the list, contrary to the initial assumption (though it seems that not everyone experiences an "aha" reaction to the proof). So this proof seems to score high on the unexpectedness parameter. It does not, however, seem to be particularly explanatory, as is often (though not always) the case with proofs by contradiction, and so it would be an example of a proof that is deemed beautiful but not chiefly in virtue of its explanatoriness.

4 Hardy's Six and explanatoriness

Having offered further support for the legitimacy of Hardy's Six as an explication of the beauty of a mathematical proof by means of a concrete example, I now argue that all but one of the six features are closely related to the notion of explanatoriness. This means that the purported non-aesthetic, purely epistemic properties that aesthetic vocabulary is tracking according to Rota and Todd (in his earlier work) might all be chiefly connected to the ideal of a proof as an explanatory device.

The distinction between nonexplanatory and explanatory mathematical proofs is often formulated in terms of the difference between proofs that merely establish *that* the conclusion is the case and proofs that establish *why* the conclusion is the case; the former merely demonstrate, while the latter *explain*. The issue of explanatoriness has received significant attention in recent decades, both among philosophers (Mancosu and Pincock, 2012) and among mathematics educators (Hanna et al., 2010), and a number of different accounts of the explanatoriness of mathematical proofs have been proposed. In the philosophical literature, two influential accounts are those proposed by Steiner (1978) and Kitcher (1989). (In Dutilh Novaes (2018) I offer my own dialogical account of the explanatoriness of mathematical proofs, which is based on the observation that a proof is a piece of discourse aimed at an intended audience, with the intent to produce explanatory persuasion. See below for more on explanatory persuasion.)

Steiner introduced the notion of a *characterizing property* as what distinguishes explanatory from nonexplanatory proofs: "An explanatory proof makes reference to a characterizing property of an entity or structure mentioned in the theorem, such that from the proof it is evident that the result depends on the property" (Steiner, 1978: 143). This notion of dependence seeks to capture a noncausal but realist analogue of the notion of causation in scientific explanation more generally. As described by Mancosu and Pincock, "Steiner argues that these dependence relations require both that the entity or structure be uniquely picked out by some characterizing property, and that the explanatory proof be part of a family of proofs where this property is varied" (Mancosu and Pincock, 2012: 16). Importantly, the notion of a characterizing property is attributed to mathematical entities or structures, and is thus conceived of in realistic, objective terms. However, there is still an irreducible epistemic component in how the characterizing property is (or is not) suitably captured in the proof.

As pointed out by Mancosu (2011), Steiner's is a local account, that is, explanatoriness is understood as a local property of a given proof. In contrast, Kitcher's (1989) approach is aptly described as global/holistic in that the explanatoriness of proofs is viewed in the broader context of (mathematical) knowledge as a whole. For Kitcher, the key notion is that of *unification*: "An explanatory proof in pure mathematics is one that is part of a small collection of argument patterns that allows the derivation of the mathematical claims that we accept" (Mancosu and Pincock, 2012: 15).

It is clear that Hardy's Six bear close connections with the ideal of explanatoriness for proofs, thus understood. In particular, generality, inevitability, and simplicity are consequences of Steiner's local account of explanatoriness. Generality ensues from variations of the characterizing property in the relevant family of proofs; inevitability follows from it being evident from the proof that the result depends on the relevant characterizing property; simplicity is guaranteed by the central role in the proof of one single characterizing property. In turn, depth and especially seriousness are related to Kitcher's holistic conception of explanatoriness. If a proof is deep and thus deals with foundational concepts/mathematical entities, then it will likely be instrumental in the derivation of many widely accepted mathematical claims. If it is serious and thus connected to many other mathematical ideas, then it is well placed to play the unifying role that Kitcher attributes to explanatory proofs.[22]

In sum, five features among Hardy's Six bear close connections with properties that have been associated with explanatoriness in the philosophical literature. But what about *unexpectedness*? Is it also a feature in a proof that enhances explanatoriness? I submit that the opposite is the case, in fact. In an explanatory proof, there should be no surprises: each step in the proof must be clear and evident, eliciting immediate understanding in whoever inspects the proof, thus ruling out unexpected "turns."[23] In an explanatory proof, presumably we know all along where we are coming from and where we are going; such a proof is epistemically transparent throughout, and so there should be no trace of unexpectedness in it.[24]

But if, as I claim, unexpectedness does not enhance the explanatoriness of a proof, then why do mathematicians attribute a positive value to unexpectedness? (That is, if we accept Hardy's claim that unexpectedness is one of the features that make a proof beautiful.) What role does surprise and unexpectedness have to play in the practice of mathematical proofs, if not related to enhancing explanatoriness? In order to address this question, a short detour is required to introduce a dialogical conception of mathematical proof.

5 Dialogical conception: Explanatory persuasion

Within the (broadly conceived) practice-based approach adopted here, one pressing question becomes: What is the point of a proof? Why do mathematicians bother producing proofs at all? What effect(s) do they seek to obtain? How is the practice of proofs related to the more general enterprise of producing and sharing mathematical knowledge? It is crucial that we address these questions in the context of a thorough investigation of practices of attributing aesthetic properties to proofs. Mathematicians and philosophers of mathematics tend to overlook these important questions, but they have been addressed by at least a few prominent authors (Hersh, 1993; Rav, 1999; Dawson, 2006; Auslander, 2008).

One promising vantage point to address these questions is what could be described as a *genealogical approach* (Dutilh Novaes, 2015): it makes sense to inquire into what the first practitioners of a given practice thought they were doing, and why they were doing it when the practice first emerged. (But note the likelihood of *shifts of function* along the way.) In this case, the historical emergence of deductive proof in ancient Greek mathematics is a particularly relevant data point, and on this topic the most authoritative study remains (Netz, 1999). Netz emphasizes the importance of persuasion, orality, and dialogue for the emergence of classical, "Euclidean" mathematics in ancient Greece:

> Greek mathematics reflects the importance of *persuasion*. It reflects the role of orality, in the use of formulae, in the structure of proofs, and in its reference to an immediately present visual object. But this orality is regimented into a written form, where vocabulary is limited, presentations follow a relatively rigid pattern, and the immediate object is transformed into the written diagram—doubly written, for it is now inscribed with letters . . . . It is at once oral and written. (Netz, 1999: 297–98)

Netz's interpretation relies on earlier work by Lloyd (1996), who argues that the social, cultural, and political context in ancient Greece, and in particular the role of practices of *debating*, was fundamental for the emergence of the technique of mathematical deductive proofs (see also (Jahnke, 2010)). So from this perspective, it seems that one of the main functions of deductive proofs (then as well as now) is to produce *persuasion*, in particular what one could call *explanatory persuasion*: to show not only *that* something is the case, but also *why* it is the case.[25] As well put by Dawson,

> We shall take a proof to be an informal argument whose purpose is to convince those who endeavor to follow it that a certain mathematical statement is true (and, ideally, to explain *why* it is true). (Dawson, 2006: 270)

What I add to Dawson's description is an explicit multi-agent, dialogical perspective (as suggested by Netz's historical analysis), which is left implicit in his description. On this conception, a deductive proof corresponds to a dialogue between the person wishing to establish the conclusion (given the presumed truth of the premises), and an interlocutor who will not be easily convinced and who will bring up objections, counterexamples, and requests for further clarification and precision. A good proof is one that convinces a fair but "tough" opponent; as (allegedly) noted by the mathematician Mark Kac, "the beauty of a mathematical proof is that it convinces even a stubborn proponent" (Fisher, 1989: 50). (Notice yet another reference to the aesthetic dimension of proofs, this time connected to persuasion.) Now, if this is right, then mathematical proof is an inherently dialogical, multi-agent notion, given that it is essentially a piece of discourse aimed at a putative audience, typically composed of "stubborn" interlocutors.

To be sure, there are different ways in which the claim that mathematical proofs are essentially dialogical can be understood. For example, the fact that a proof is only recognized as such by the mathematical community once it has been sufficiently scrutinized by trustworthy experts can also be viewed as a dialogical phenomenon, perhaps in a loose sense (the "dialogue" between the mathematician who formulates a proof and the mathematical community who scrutinizes it).[26] But in what follows I present a more precise rational reconstruction of the (quite specialized) dialogues that would correspond to deductive proofs.

On this conception, proofs are semi-adversarial dialogues of a special kind involving two participants: Prover and Skeptic.[27] Prima facie, the participants have opposite goals, and this is why the adversarial component remains prominent: Prover wants to establish the truth of the conclusion, and Skeptic will not be easily convinced. The dialogue starts with Prover asking Skeptic to grant certain premises. Prover then puts forward further statements, which purportedly follow from what has been granted. (Prover may also ask Skeptic to grant additional auxiliary premises along the way.)

It may seem that all the work is done by Prover, but Skeptic has an important role to play, namely to ensure that the proof is persuasive, perspicuous, and correct.[28] Skeptic's moves are: granting premises so as to get the proof going; offering a counterexample when an inferential move by Prover is not really necessarily truth-preserving (or a global counterexample to the whole proof);[29] asking for further clarification—why-questions—when a particular inferential step by Prover is not sufficiently compelling and clear. These three moves correspond neatly to what are arguably the three main features of a mathematical proof: it starts off with certain premises; it proceeds through necessarily truth-preserving inferential steps; these steps should be individually evident and explanatory.

From this point of view, a mathematical proof is characterized by a complex interplay between adversariality and cooperation: the participants have opposite goals and "compete" with one another at a lower level, but they are also engaging in a common project to investigate the truth or falsity of a given conclusion (given the presumed truth of the premises) in a way that is not only persuasive but also (ideally) explanatory.[30] If both participants perform to the best of their abilities, then the common goal of producing mathematical knowledge will be optimally achieved.[31] The cooperative component is also confirmed by the observations that Prover seeks to assist Skeptic in truly understanding why the conclusion follows from the premises, and that (by asking appropriate why-questions) Skeptic may help Prover to formulate an explanatory proof.[32]

At this point, the reader may be wondering: this is all very well, but obviously deductive proofs are not really dialogues, as they are typically presented in writing rather than produced orally.[33] If at all, there is only one "voice" that we hear: that of Prover. So at best, they must be viewed as monologues. But a mathematical proof is arguably an instance of what linguist E. Pascual describes as a *fictive interaction*, that is, a piece of discourse that is apparently monological but in fact reproduces a multi-agent communicative scenario.

> Specifically, we present the premise that there is a conversational basis for language, which serves to partly structure cognition, discourse, and grammar. Stemming from this tenet, we discuss the notion of fictive interaction or "FI," namely the use of the template of face-to-face interaction as a cognitive domain that partially models: (i) thought (e.g. talking to oneself); (ii) the conceptualization of experience (e.g. "A long walk is the answer to headache"); (iii) *discourse organization (e.g. monologues structured as dialogues)*; and (iv) the language system and its use (e.g. rhetorical questions). (Pascual and Oakley, 2017: 348, emphasis added)

Given the observation that a mathematical proof is a (specific, regimented, and specialized) form of discourse, and the idea that interactive, conversational structures permeate much of what appears to be monological discourse at first sight, it is in fact not surprising that a mathematical proof too would have a conversational (dialogical) basis. My main contribution here is to spell out in more detail the specific kind of conversation that a mathematical proof re-enacts: semi-adversarial dialogues following fairly strict rules, involving the fictive participants Prover and Skeptic.

Indeed, Skeptic may have been "silenced," but he is still alive and well insofar as the deductive method has internalized the role of Skeptic by making it constitutive of the deductive method as such. Recall that the job of Skeptic is to look for counterexamples and to make sure the argumentation is perspicuous.[34] This in turn corresponds to the requirements that each inferential step in a proof be necessarily truth-preserving (and so immune to counterexamples), and that each step of a proof be evident and persuasive.

Further empirical support for the claim that a mathematical proof is best viewed as a fictive dialogue is the work by Hodds and colleagues on self-explanation training to increase students' proof comprehension (Hodds et al., 2014). The training consists in teaching students to ask themselves questions about proofs that are very similar to the questions that Skeptic asks: Do I understand the ideas used in this inferential step? Do I understand the general idea of the proof? Does the information provided in the proof contradict my beliefs on the topic thus far? (which may prompt a search for counterexamples.) Students are even instructed to provide answers to these questions out loud, as if engaging in a real, oral dialogue, thus enacting as it were the fictive interaction of the written form. Their results indicate that this approach substantially improves proof comprehension, thus suggesting that "going back" to a dialogical version of the proof has a significant cognitive impact. (Recall Netz's description of mathematical proofs above as "at once oral and written.")

6 The aesthetic import of unexpectedness

Given this conception of proofs as written discourse with an oral, dialogical origin, I submit that literary genres such as poetry and narrative offer a suitable perspective for an analysis of the aesthetic dimension of mathematical proofs. Of course, and as emphasized by Netz (1999) and others, visual elements such as diagrams and special notations are also widely present in mathematical proofs, and may well be a crucial component for their aesthetic appreciation.[35] But above all, mathematical proofs as currently produced and consumed seem to belong primarily to the category of *texts*. (Not coincidentally, even in classroom oral presentations, writing (e.g., on a whiteboard) typically plays an essential role.) This is indeed what Netz (2005, 2009) does, by taking the concepts of narrative structure of prose and prosodic structure in poetry as guiding threads for his investigation of the aesthetic dimensions of geometric proofs in ancient Greek mathematics. But it is not only with respect to ancient Greek mathematics that these concepts lead to fruitful analysis. For example, there is some interesting recent work on the connections between proofs and narratives (Doxiadis and Mazur, 2012).

Perhaps more relevant for the present purposes is the observation that mathematical proofs display interesting similarities with poetry.[36] In a number of poetic traditions (Greek, Chinese, Arabic, French), what is usually referred to as "classical poetry" is characterized by fairly strict sets of rules. Take the poetic form *sonnet*, for example, which originated in the court of Emperor Frederick II in Sicily (first half of the thirteenth century), and was later made famous by Petrach and exported to France, England, and beyond. (In the nineteenth-century French poetry, the sonnet was still popular, for example, with Baudelaire.) As is well-known, a sonnet is a poem of fourteen lines that follows a strict rhyme scheme and specific structure, though conventions associated with the format have evolved over its history. These conventions are not entirely arbitrary though, and indeed often serve as "recipes" for constructing interesting, unusual sounds.

In classical poetry thus understood (i.e., as following fairly strict rules for rhyme, number of syllables, prosody etc.), one of the main sources of poetic beauty is precisely the phenomenon of *creativity within constraints*: the tension between the constraints imposed by the rules and how the poet manages to discover innovative and unexpected solutions to these constraints. It is the difference between, say, a dull sonnet that perfectly complies with the rules for the format and yet fails to "ignite a spark," and an inspired sonnet that goes much beyond simply complying with the rules by introducing unexpected, appealing, novel combinations of sounds and meanings. This general point applies to other artistic forms where more or less formal constraints apply, for example, classical music: it is often unexpected combinations that differentiate a masterpiece from a formally adequate but dull piece.[37]

I submit that the aesthetic import of unexpectedness with respect to mathematical proofs identified by Hardy is similar to its aesthetic import in artistic genres such as classical poetry and classical music: a creative solution to a set of constraints provokes pleasurable feelings.[38] This parallel is aptly captured in the following passage by Fields-medalist Timothy Gowers in his *Mathematics: A Very Short Introduction*.

> It is a notable feature of the second argument[39] that it depends on a single idea, which, though unexpected, seems very natural as soon as one has understood it. It often puzzles people when mathematicians use words like "elegant," "beautiful," or even "witty" to describe proofs, but an example such as this gives an idea of what they mean. Music provides a useful analogy: we may be entranced when a piece moves in an unexpected harmonic direction that later comes to seem wonderfully appropriate, or when an orchestral texture appears to be more than the sum of its parts in a way that we do not fully understand. Mathematical proofs can provide a similar pleasure with sudden revelations, unexpected yet natural ideas, and intriguing hints that there is more to be discovered. (Gowers, 2002: 51)[40]

Gowers' remarks are illuminating, and point in a promising direction: unexpectedness in mathematical proofs can cause pleasurable aesthetic feelings, comparable to unexpectedness in music. However, these remarks by themselves cannot be considered as decisive arguments on the issue of the role of unexpectedness in cognitive processes pertaining to proofs. (Recall the methodological considerations on the limited evidentiary role of self-reported and anecdotal data discussed under "Overview of literature.") Fortunately, the role of unexpectedness in cognitive processes has been investigated empirically by cognitive scientists, so we can approach the issue in a more systematic way. In particular, the connection between *unexpectedness* and the *persuasiveness* of a proof will be examined.

7 Unexpectedness and the role of affective responses

While there may be much in common between experiences of aesthetic pleasure related to unexpectedness in music, poetry, and mathematics, there is still an important distinction between predominantly artistic enterprises such as music and poetry, and predominantly epistemic/cognitive enterprises such as mathematics. In particular, we may ask ourselves how experiences of aesthetic pleasure or displeasure affect the epistemic/cognitive components of mathematical activity, a question that is less pressing with respect to artistic domains.[41] In this section, I argue that aesthetic pleasure, in particular but not exclusively pleasure resulting from unexpectedness, increases the mathematician's conviction in the proof in question, thus increasing its persuasiveness, by eliciting positive affective responses toward the proof. In the same vein, "ugly" proofs are perceived as less convincing even if formally correct.[42] The conclusion will be that affective responses, be they positive or negative, are an integral part of "doing mathematics," and this is arguably part of what is being tracked by the aesthetic vocabulary used by mathematicians to describe proofs.

Given the crude but still widespread opposition between reason and emotions, it might be expected that in mathematics, which is often seen as the pinnacle of reason, emotions should have no role to play whatsoever. However, over the last decades, a number of influential researchers have questioned this crude opposition; a seminal work in this tradition is Damasio's *Descartes' Error* (Damasio, 1994). Indeed, there is now a substantive body of research strongly suggesting that emotions and affective responses are inextricable components of human cognitive and epistemic processes (Reber, 2016).

Indeed, it follows quite straightforwardly from the dialogical account of mathematical proof presented above that affective responses should play at least some role in the practice of mathematical proofs, given the centrality of the concept of *persuasion*. However, the role of emotions and affective responses with respect to mathematical knowledge is yet to be more extensively investigated.

There is some work by mathematics educators on the role of affective responses in mathematical problem solving (McLeod and Adams, 1989), and some work by cognitive scientists on intuitions in mathematics (Reber et al., 2008), but for the most part it remains terrain to be further explored.

For our purposes, the work of R. Reber and colleagues is particularly relevant. Reber is known for his work on processing fluency, and in particular for his processing fluency theory of aesthetic pleasure. The key idea is that a receiver experiences aesthetic pleasure from the fact that an object (a piece of music, an argument, visual stimulus) can be processed with ease (Reber et al., 2004). Processing fluency has been shown to influence affect (Reber et al., 1998) and judgments of truth (Reber and Unkelbach, 2010). Especially relevant for our purposes are so-called aha-experiences, which are typically described as characterized by "suddenness (the experience is surprising and immediate), ease (the solution is processed without difficulty), positive affect (insights are gratifying), and the feeling of being right (after an insight, problem solvers judge the solution as being true and have confidence in this judgment)" (Topolinski and Reber, 2010: 402). Topolinski and Reber propose an account of aha-experiences in terms of fluency:

> We propose a fluency account of insight: Positive affect and perceived truth and confidence in one's own judgment are triggered by the sudden appearance of the solution for a problem and the concomitant surprising fluency gain in processing. We relate earlier evidence on insight concerning the impact of sudden fluency variations on positive affect and perceived truth and confidence. (Topolinski and Reber, 2010: 402)

This means that, when the unexpectedness of an argumentative strategy in a mathematical proof leads to an aha-experience (the sudden appearance of the solution for the problem), positive affect and accompanying perceived truth/confidence in the result are triggered or enhanced. In other words, a surprising mathematical proof has two interrelated effects: it elicits positive affective responses, which in turn increase the feeling of "being right." Now, at least some of these positive affective responses are quite naturally interpreted as aesthetic experiences (Silver and Metzger, 1989), and moreover they increase the persuasiveness of the proof in the eyes of its receiver.[43] This might explain why mathematicians often deem surprising proofs to be more convincing and more beautiful; arguably, this is a result of the pleasurable feelings arising from the accompanying aha-experience, not of purely non-affective epistemic processes (e.g., those related to explanatoriness).

We can now return to Cantor's diagonal argument, which illustrates these general points quite vividly. The proof is arguably not particularly explanatory, and many people fail to see at a glance that number D* cannot be on the list of decimal expansions between 0 and 1 (some people voice the suspicion that there must be some "trick" going on). So presumably, the pleasurable aesthetic feelings provoked by the proof do not arise only from its non-affective epistemic dimension related to explanatoriness. Instead, it seems that it is the "plot twist" of the appearance of a number that should have been on the list and yet cannot be, and the accompanying aha-experience when one realizes that it really cannot be on the list (given how it was generated), that lead to a positive aesthetic assessment of the proof as beautiful. Tellingly, those who do not undergo the relevant aha-experience (and thus do not experience the pleasurable feeling accompanying it) tend to find the proof unconvincing even when they cannot quite pinpoint what exactly is wrong with it. Cantor's proof is thus a clear example of a proof where unexpectedness, persuasiveness, and beauty all come together.

These considerations suggest that, when it comes to mathematical proofs (as with cognitive processes more generally), epistemic and affective components are inherently intertwined. Moreover, since at least some of these affective reactions are plausibly interpreted as aesthetic experiences, this in turn suggests that the alleged opposition between the aesthetic and the epistemic in mathematical proofs is misguided. Here, I've focused on unexpectedness as an aesthetic-epistemic component of proofs, related in particular to proofs as pieces of discourse aimed at persuasion, but there may well be other such aesthetic-epistemic components.

Todd (2018) reaches similar conclusions, even though he rejects Reber's account of aha-experiences in terms of fluency. (I agree with Todd that the account is not entirely convincing.) Instead, Todd develops the notion of "aesthetic-epistemic feelings," drawing on the notion of epistemic feelings (Arango-Muñoz, 2014). Roughly, epistemic feelings can be described as "the valenced experienced manifestation of some kind of quick and dirty, low-level, sub-personal mechanism that monitors the performance of different cognitive processes" (Todd, 2018: 11). Todd then goes on to argue that, given significant similarities between epistemic feelings thus understood and a range of aesthetic feelings (valence, quick and dirty responses, relevance of features such as symmetry, simplicity etc.), there does not seem to be a clear-cut distinction between them; perhaps at least some of them can be clustered under the common concept of "aesthetic-epistemic feelings."

So, I contend, we can think of (some) epistemic feelings as possessing aesthetic attributes, or of (some) aesthetic feelings as exhibiting epistemic attributes, or of a range of feelings that are jointly aesthetic-epistemic in nature. . . . I will suggest that, at least from an explanatory point of view there is little to choose between [these options] and they may in fact be perfectly compatible. (Todd, 2018: 17–18)

To be clear, Todd is not contending that there is no difference whatsoever between epistemic and aesthetic feelings in general; rather, he is claiming that there is an interesting subclass of feelings that seem to be both. Furthermore, he claims that the prima facie aesthetic assessments of mathematical proofs pertain to phenomena falling under this category. The conclusion is thus again that the epistemic versus the aesthetic opposition when it comes to mathematical proofs is a false dichotomy.[44]

8 Empirical predictions

These observations give rise to fairly precise empirical predictions that can be put to test, for example by using the methodology developed by Inglis and Aberdein (for some of these predictions, at least some preliminary results are already available).

Beauty and explanatoriness are used as comparative, graded notions. The claim is that these are notions predominantly used to express predilections for certain proofs over others, and thus are essentially comparative. The existing results by Inglis and Aberdein seem to support this prediction, but they have not addressed this question specifically by, for example, asking participants to *compare* two different proofs on a number of parameters.

Judgments of beauty and explanatoriness correlate, but do not coincide. This follows from the idea that the beauty (or ugliness) of a proof can be largely but not entirely understood in epistemic terms related to explanatoriness. In the results reported in Inglis and Aberdein (2015), the correlation between beauty and explanatoriness is greater than zero, but it is not that high, at only .20 (Table 4 on p. 101)—admittedly, lower than the present account would predict. However, "beautiful" correlates strongly with "enlightening" at .57, and I've argued above that the property of enlightenment (as described by Rota) is very closely related to that of explanatoriness. So here too we need further research, for example, with a more graded scale of appraisal, to establish the extent to which mathematicians tend to view what they consider to be explanatory proofs as also beautiful. Some interesting questions are whether a proof can be viewed as beautiful without being viewed as explanatory, or whether two proofs may be deemed equally explanatory but not equally beautiful, which would indicate a weaker connection between beauty and explanatoriness.

Judgments of beauty and unexpectedness/creativity strongly correlate. This follows from the claim that, just as in other domains such as classical music and classical poetry, in mathematical proofs beauty is connected to surprising, creative solutions within a set of constraints. This is clearly borne out by the results in Inglis and Aberdein (2015), where the following adjectives correlate strongly with "beautiful" (Table 4): "ingenious" (.60), "striking" (.59), "creative" (.55), "innovative" (.42).

Judgments of beauty and persuasiveness strongly correlate. This follows from the claim that aesthetic pleasure enhances conviction. Indeed, some of the proofs that are typically viewed as "ugly" such as probabilistic or computer-assisted proofs tend to be viewed as unpersuasive as well, in part because they fail to transfer to the audience the reason(s) why the theorem must be true (Easwaran, 2009). Inglis and Aberdein (2015) did not include "persuasive" among the adjectives on their list, so at this point a systematic investigation of the correlation between attributions of beauty and attributions of persuasiveness with respect to proofs remains to be done.

Mathematicians disagree in their aesthetic judgments about proofs.[45] This follows from the idea that explanatoriness and beauty are audience-relative notions, though it may well be that certain kinds of proofs (e.g. "brute force" combinatorial proofs) are unanimously seen as ugly. Inglis and Aberdein's results on diversity in proof appraisal (2016) also suggest this much, but hitherto they only worked with a small range of proofs. Investigation of mathematicians' appraisals of a much larger range of proofs is needed for more general conclusions to be drawn on how systematic the phenomenon of diversity in proof appraisal is.

9 Conclusions

My starting point in this chapter was the pervasiveness of uses of aesthetic vocabulary to talk about mathematical proofs, a phenomenon that calls for a philosophical explanation. I have argued that, contrary to what seems to be presupposed in much of the literature on the topic, there is no real opposition between the idea that these uses are tracking aesthetic properties and the idea that they are tracking epistemic properties; the aesthetic and the epistemic overlap when it comes to mathematical proofs (as well as in other domains). In particular, positive affective responses elicited by, for example, unexpectedness in how a proof is formulated increase the mathematician's conviction in the proof. And thus, the issue of the presumed beauty of mathematical proofs leads us to the very important but still largely unexplored issue of the role(s) of affective responses and emotions in mathematics. The present chapter is but a small contribution in this direction; it is to be hoped that philosophers of mathematics, mathematics educators, and cognitive scientists working on mathematical cognition will continue to investigate the role of affect and emotions in mathematics.

Notes

1 Aesthetic properties can be attributed to mathematical objects other than proofs as well, but in the present investigation we will focus on proofs specifically.
2 Literal versus nonliteral is Montano's (2014) terminology; non-reductive versus reductive is Inglis and Aberdein's (2015) terminology.
3 For example, simplicity for McAllister (2005); or fit for Raman-Sundström (2012), which is a property of the *relation* between a theorem and its proof.
4 Another way in which the presumed dichotomy between the aesthetic and the epistemic can be dissolved is by relying on the notion of *functional beauty* (Bueno, 2009; Paterson, 2013). The basic idea would be that if the function of a proof is chiefly epistemic, then by performing its function well, a proof is beautiful precisely in virtue of its epistemic properties. I've defended this general idea in a number of talks over the years, and I still think there is much to be said in its favor. However, I have come to think that it ultimately fails to give a satisfactory account of the phenomena in question; there seems to be more to the aesthetic dimension of proofs than what is captured by the notion of functional beauty alone.
5 One reason why mathematicians hold simplicity in high esteem may be that, if mathematics progresses according to the Lakatosian model of "proofs and refutations" (Lakatos, 1976), there is a tendency for proofs to become cumbersome and convoluted when objections (refutations) are raised and adjustments are made (what Lakatos describes as "lemma incorporation"). So the ideal of simplicity (which may or may not be associated with beauty) would serve to bring counterbalance to this tendency.
6 See also (Raman-Sundström, 2012) on fit.
7 Colyvan (2012: 79) also notes that judgments of explanatoriness are typically left out of the published work of mathematicians. Furthermore, he notes, "It is difficult

to find a great deal of agreement in the mathematical literature on which proofs are explanatory and which are not." Whether mathematicians converge in their attributions of beauty and of explanatoriness to proofs is still essentially an open question. Much of the literature on both topics seems to presuppose that they do, but this is not to be taken for granted. In fact, recent studies conducted by Inglis and different sets of collaborators (Inglis et al., 2013) (Inglis and Aberdein, 2016) suggest significant divergence in mathematicians' appraisals of proofs, both on aesthetic and on epistemic dimensions.

8   "I argue that . . . there are strong reasons for suspecting that many, and perhaps all, of the supposedly aesthetic claims are not genuinely aesthetic but are in fact 'masked' epistemic assessments" (Todd, 2008: 61). Notice that, in recent work, Todd (2018) defends a more nuanced account of the relations between the aesthetic and the epistemic. Todd's recent work will be discussed later on.

9   There is of course a rather trivial sense in which the epistemic dimension of a proof is related to aesthetic considerations, namely the fact that (presumably) only those who minimally understand a proof in the first place will be able to appreciate its beauty or ugliness qua proofs (as noted in Raman-Sundström (2012)). But Todd's and Rota's reductive analyses go beyond the idea that understanding a proof is a necessary (though perhaps not sufficient) condition for the appreciation of its (purported) aesthetic properties by arguing that there is ultimately no truly aesthetic property being grasped.

10 Compare (Mancosu, 2011: section 7) on top-down approaches to the notion of mathematical explanation: "Previous theories of mathematical explanation proceeded top-down, that is by first providing a general model without much concern for describing the phenomenology from mathematical practice that the theory should account for. Recent work has shown that it might be more fruitful to proceed bottom-up, that is by first providing a good sample of case studies before proposing a single encompassing model of mathematical explanation."

11 Of course, the experiences and reports of working mathematicians *can* serve as a suitable starting point for philosophical analysis, such as in Paterson (2013). The point is rather that these testimonies may not be sufficient to settle philosophical questions pertaining to aesthetic judgments in mathematics.

12 The potential heterogeneity of the phenomenon being explained suggests that the case-study methodology advocated by Mancosu may also not be entirely satisfying; the range of case studies analyzed must be large enough so as to be representative of the potential variety of prima facie aesthetic judgments on proofs. Notice also that Mancosu himself (Hafner and Mancosu, 2005) is sympathetic to the idea of non-homogeneous phenomena as the objects of study for philosophical accounts of mathematics and mathematical practice.

13 In recent work, Inglis and Aberdein (2017) have been tackling some of these concerns. See also (Larvor, 2016).

14 But notice that nonempirical methodologies such as good-old conceptual analysis and historical analysis (e.g. case studies) remain important. Arguably, what we need is a combination of methodologies. (For an account of what I refer to as "conjunctive methodological pluralism," see (Dutilh Novaes, 2012: Conclusion).)

15 More generally, it may well be that competing accounts that differ substantially from Hardy's are equally compelling. However, of the accounts available in the literature, Hardy's seems to me to be the richest and most wide-ranging (multidimensional), and thus most suitable for the present purposes.

16 Notice that the depth criterion is closely related to Hardy's realist, platonist conception of mathematics: some mathematical concepts are inherently more fundamental, that is, "deeper," than others according to a stratified organization.

17 Naturally, if the real numbers in this small interval are uncountable, as will be proved by contradiction, then the set of all real numbers will be uncountable too.

18 We know that each real can be represented by a decimal expansion, in fact by at least 1 and at most 2 decimal expansions.

19 See (Floyd, 2012) for Wittgenstein's criticism of Cantor's proof, and (Hodges, 1998) for some failed attempts at refuting the proof.

20 Notice though that this is obviously an empirical question, one that, to my knowledge, has not been rigorously studied. This is, first of all, my own instinctive response to the proof, and of a majority of members of audiences when I gave talks on this material over the years. But interestingly, some members of these audiences did voice the opinion that they do not find this proof particularly beautiful.

21 One natural question that arises is whether proofs can be attributed aesthetic properties other than beautiful, elegant, ugly, etc. For example, can a proof be *funny*? I do not have the space to discuss this fascinating question, but my quick answer is: yes, it can. An example would be *Hilbert's Hotel*, a humorous adaptation of the gist of Cantor's diagonal argument (Bellos, 2010: chapter 11).

22 Notice that the order of entailments here is different: Steiner's notion of explanatoriness entails generality, inevitability, and simplicity; depth and seriousness entail Kitcher's notion of explanatoriness as unification.

23 Notice that the surprise referred to here, which I claim explanatory proofs typically do not display, does not pertain to the conclusion itself, but rather to the *path* leading from premises to conclusion. Indeed, there can be a surprising proof of a theorem that was nevertheless widely held to be true even before the proof was produced, or a non-surprising proof of a surprising result (though that's probably less likely to happen).

24 For discussion of the inverse relation between surprise and explanatoriness (explanation decreases surprise), see (Schupbach and Sprenger, 2011).

25 For Hersh (1993), proof is also about convincing and explaining, but on his account these two aspects come apart. According to him, convincing is aimed at one's

mathematical peers, while explaining is relevant in particular in the context of teaching. Similarly, in argumentation theory, persuasion and understanding are often thought to be orthogonal phenomena: persuasion would be related to a mild form of coercion via argumentation—one cannot *but* assent to the conclusion—whereas understanding via explanation presupposes a certain amount of cognitive freedom. (See (Wright, 1990) on the two phenomena.) On my account, however, persuasion and explanation go hand in hand in mathematical proofs; a proof will be more persuasive precisely if it is viewed as (more) explanatory.

26 Take, for example, the saga of Mochizuki's purported proof of the ABC conjecture, which remains impenetrable for the mathematical community at large, and so has been in limbo for years (Castelvecchi, 2015). See also (Auslander, 2008) for proof as certification.

27 This terminology comes from the computer science literature on proofs. I borrow it from (Sørensen and Urzyczyn, 2006), who speak of *prover-skeptic games*, but the terminology seems to have been "in the air" for a while (Kolata, 1986). One may think of the interplay between proofs and refutations as described in Lakatos's *Proofs and Refutations* as an illustration of this general idea: Prover aims at proofs, Skeptic aims at refutations. The "semi" qualification pertains to the equally strong cooperative component in a proof, to be discussed shortly. See also (Pease et al., 2017) for a formalized version of Lakatosian games of proofs and refutations, again very much in the spirit of the Prover-Skeptic dialogues here described.

28 Moreover, again on a Lakatosian picture, refutations and counterexamples brought up by Skeptic may play the fundamental role of refining the conjectures and their proofs.

29 Lakatos (1976) distinguishes between global and local counterexamples.

30 As well put by one of the characters in *Proofs and Refutations*: "Then not only do refutations act as fermenting agents for proof-analysis, but proof-analysis may act as a fermenting agent for refutations! What an unholy alliance between seeming enemies!" (Lakatos, 1976: 48). See also recent discussions on "adversarial collaboration" in the social sciences (Mellers, Hertwig, and Kahneman, 2001).

31 Compare to what happens in a court of law in adversarial justice systems: defense and prosecution are defending different viewpoints, and thus in some sense competing with one another, but the ultimate common goal is to achieve justice. The presupposition is that justice will be best served if all parties perform to the best of their abilities.

32 The cooperative component becomes immediately apparent if one considers that a one-line "proof" from premises to conclusion—say, from the axioms of number theory straight to Fermat's Last Theorem—will be necessarily truth-preserving, and yet will not count as an adequate proof. But of course, Skeptic may also make misuse of why-questions and refuse to be convinced even when a particular

inferential step is as clear as it can get (such as the tortoise in L. Carroll's famous story of Achilles and the Tortoise).

33 Though of course they can also be presented orally, for example in the context of teaching, but even then writing also typically occurs.

34 The work of L. Andersen (2017) interviewing mathematicians on their refereeing practices shows that, when refereeing a paper, mathematicians behave very much like the fictive Skeptic. This suggests that, in the broader social context of mathematical practices, Skeptic does remain active insofar as this role is played by members of the community who scrutinize proofs (in particular, but not exclusively, in their capacity as referees).

35 See (Cellucci, 2015) on the role of perceptual features in the appreciation of mathematical beauty.

36 The similarities between poetry and (my conception of) mathematical proofs were first pointed out to me by R. A. Briggs (personal communication), who is not only a mathematically trained philosopher but also an accomplished poet.

37 See (Judge, 2018) on musical surprise.

38 This observation is aligned with the Kantian account of the beauty of mathematical proofs proposed in Breitenbach (2013), in particular the emphasis on creativity. See also Levinson (2006) on Jon Elster's well-known account of artistic creativity as optimizing choice within constraints.

39 Gowers is comparing two proofs on the well-known problem of tiling a square grid with the corners removed (answer: it can't be done). "This short argument illustrates very well how a proof can offer more than just a guarantee that a statement is true. ... Both of them establish what we want, but only the second gives us anything like a *reason* for the tiling being impossible" (Gowers, 2002: 50–51). In the terminology we've been using, the second proof is more explanatory than the first one because it gives us a reason, an answer to the relevant why-question.

40 This passage is also cited in Todd (2018).

41 However, aesthetically pleasurable experiences in these different domains have in common that they all typically require substantive intellectual engagement, usually in terms of extensive intellectual training required to appreciate the beauty of the objects in question (poems, pieces of music, mathematical proofs).

42 Proofs described as "ugly" are often those that are considered nonexplanatory, but also those that proceed by "brute force," in mechanical ways without recourse to ingenuity (Montaño 2012). In such proofs, there is not much room for surprises, which is (I submit) one of the reasons why mathematicians dislike them.

43 That affect is related to persuasion is of course not in any way a surprising claim; this connection has been investigated for millennia by those interested in rhetoric, and more recently by marketing and business experts such as R. Cialdini (1984).

44 But notice that the interaction of epistemic and aesthetic factors can also have not-so-pretty implications. (I owe this point to S. Killmister.) Aesthetic considerations such as simplicity and elegance can in fact also lead mathematicians and other theorists astray, epistemically speaking. A recent example would be the infatuation of economists with beautiful, simple mathematical models of economical phenomena, which however were empirically wholly inadequate; this is viewed as one of the factors leading to the 2008 financial crisis. It is fair to say that in this case, theorists were dazzled by the beauty of their models in ways that distracted them from epistemic concerns such as accuracy, predictability, etc. In the same vein, a recent book with the revealing title *Lost in Math: How Beauty Leads Physics Astray* (Hossenfelder, 2018) details how aesthetic considerations seem to be hindering rather than facilitating progress in physics.

45 In this chapter, I did not discuss the issue of convergence of judgments in detail. But the claim that mathematicians should disagree in their attributions of aesthetic properties to proofs follows from the claim that much of these judgments are in fact tracking epistemic properties such as explanatoriness, and that these epistemic properties are agent-relative (as I argued in Dutilh Novaes (2018)).

References

Andersen, L. (2017). *Social Epistemology and Mathematical Practice: Dependence, Peer Review, and Joint Commitments.* Aarhus: University of Aarhus, PhD thesis.

Arango-Muñoz, S. (2014). The nature of epistemic feelings. *Philosophical Psychology*, 27:193–211.

Auslander, J. (2008). On the roles of proof in mathematics. In B. Gold and R. Simons (Eds.), *Proof and Other Dilemmas: Mathematics and Philosophy*, 61–77. Washington DC: Mathematical Association of America.

Bellos, A. (2010). *Alex's Adventures in Numberland*, London: Bloomsbury Publishing.

Breitenbach, A. (2013). Beauty in proofs: Kant on aesthetics in mathematics. *European Journal of Philosophy*, 23:955–77.

Bueno, O. (2009). Functional beauty: Some applications, some worries. *Philosophical Books*, 50:47–54.

Castelvecchi, D. (2015). The impenetrable proof. *Nature*, 526:178–81.

Cellucci, C. (2015). Mathematical beauty, understanding, and discovery. *Foundations of Science*, 20:339–55.

Cialdini, R. B. (1984). *Influence: The Psychology of Persuasion*. New York, NY: Harper Collins.

Colyvan, M. (2012). *Introduction to the Philosophy of Mathematics*. Cambridge: Cambridge University Press.

Damasio, A. (1994). *Descartes' Error: Emotion, Reason, and the Human Brain*. New York, NY: Putnam Publishing.
Dawson Jr, J. W. (2006). Why do mathematicians re-prove theorems? *Philosophia Mathematica*, 14:269–86.
Doxiadis, A. and Mazur, B. (2012). *Circles Disturbed: The Interplay of Mathematics and Narrative*. Princeton, NJ: Princeton University Press.
Dutilh Novaes, C. (2012). *Formal Languages in Logic – A Philosophical and Cognitive Analysis*, Cambridge: Cambridge University Press.
Dutilh Novaes, C. (2013). A dialogical account of deductive reasoning as a case study for how culture shapes cognition. *Journal of Cognition and Culture*, 13:459–82.
Dutilh Novaes, C. (2015). Conceptual genealogy for analytic philosophy. In J. Bell, A. Cutrofello and P. Livingston (Red.), *Beyond the Analytic-Continental Divide: Pluralist Philosophy in the Twenty-First Century*, 75–108. London: Routledge.
Dutilh Novaes, C. (2016). Reductio ad absurdum from a dialogical perspective. *Philosophical Studies*, 173:2605–28.
Dutilh Novaes, C. (2018). A dialogical conception of explanation in mathematical proofs. In P. Ernest (Ed.), *Philosophy of Mathematics Education Today*, 81–98. Berlin: Springer.
Easwaran, K. (2009). Probabilistic proofs and transferability. *Philosophia Mathematica*, 17:341–62.
Fisher, M. (1989). Phases and phase diagrams: Gibbs' legacy today. In G. Mostow and D. Caldi (Eds.), *Proceedings of the Gibbs Symposium: Yale University, May 15–17*, 39–72. Washington, DC: American Mathematical Society.
Floyd, J. (2012). Wittgenstein's diagonal argument: A variation on Cantor and Turing. In P. Dybjer, S. Lindström, E. Palmgren and B. G. Sundholm (Eds.), *Epistemology versus Ontology*, 25–44. Dordrecht, The Netherlands: Springer.
Gowers, T. (2002). *Mathematics: A Very Short Introduction*. Oxford: Oxford University Press.
Hafner, J. and Mancosu, P. (2005). The varieties of mathematical explanation. In P. Mancosu, *Visualization, Explanation and Reasoning Styles in Mathematics*, 215–50. Berlin: Springer.
Hanna, G., Jahnke, H. N. and Pulte, H. (2010). *Explanation and Proof in Mathematics – Philosophical and Educational Perspectives*. Berlin: Springer.
Hardy, G. (1940). *A Mathematician's Apology*. Cambridge: Cambridge University Press.
Hersh, R. (1993). Proving is convincing and explaining. *Educational Studies in Mathematics*, 24:389–99.
Hodds, M., Alcock, L. and Inglis, M. (2014). Self-explanation training improves proof comprehension. *Journal for Research in Mathematics Education*, 45:62–101.
Hodges, W. (1998). An editor recalls some hopeless papers. *Bulletin of Symbolic Logic*, 4:1–16.
Holden, H. and Piene, R. (2009). *The Abel Prize 2003–2007: The First Five Years*, Heidelberg: Springer.

Hossenfelder, S. (2018). *Lost in Math: How Beauty Leads Physics Astray*. New York, NY: Basic Books.
Inglis, M. and Aberdein, A. (2015). Beauty is not simplicity: An analysis of mathematicians' proof appraisals. *Philosophia Mathematica*, 23:87–109.
Inglis, M. and Aberdein, A. (2016). Diversity in proof appraisal. In B. Larvor, *Mathematical Cultures: The London Meetings 2012-2014*, 163–79. Basel: Birkhäuser.
Inglis, M. and Aberdein, A. (2017). Social Influence and Conformity in Mathematicians' Aesthetic Judgements. *Manuscript*.
Inglis, M., Mejía-Ramos, J. P., Weber, K. and Alcock, L. (2013). On mathematicians' different standards when evaluating elementary proofs. *Topics in Cognitive Science*, 5:270–82.
Jahnke, H. N. (2010). The conjoint origin of proof and theoretical physics. In G. Hanna, H. Jahnke and H. Pulte (Eds.), *Explanation and Proof in Mathematics*, 17–32. New York, NY: Springer.
Judge, J. (2018). The surprising thing about musical surprise. *Analysis*, 78(2):225–34.
Kitcher, P. (1989). Explanatory unification and the causal structure of the world. In P. Kitcher and W. Salmon (Eds.), *Scientific Explanation*, 410–505. Minneapolis: University of Minnesota Press.
Kolata, G. (1986). Prime tests and keeping proofs secret. *Science*, 233:938–39.
Lakatos, I. (1976). *Proofs and Refutations*. Cambridge: Cambridge University Press.
Larvor, B. (2016). What are mathematical cultures? In S. Ju, B. Löwe, T. Müller and Y. Xie (Eds.), *Cultures of Mathematics and Logic: Selected Papers from the Conference in Guangzhou, China, November 9–12, 2012*, 1–22. Basel: Birkhäuser.
Levinson, J. (2006). Elster on artistic creativity. In *Contemplating Art: Essays in Aesthetics*, 56–76. Oxford: Oxford University Press.
Lloyd, G. (1996). Science in antiquity: The Greek and Chinese cases and their relevance to the problem of culture and cognition. In D. Olson and N. Torrance (Eds.), *Modes of Thought: Explorations in Culture and Cognition*, 15–33. Cambridge: Cambridge University Press.
Mancosu, P. (2011). Explanation in mathematics. In E. Zalta (Ed.), *Stanford Encyclopedia of Philosophy*. https://plato.stanford.edu/entries/mathematics-explanation/
Mancosu, P. and Pincock, C. (2012). *Mathematical Explanation*, Oxford Bibliographies. Oxford: Oxford University Press.
McAllister, J. (2005). Mathematical beauty and the evolution of the standards of mathematical proof. In M. Emmer (Eds.), *The Visual Mind II*, 15–34. Cambridge, MA: MIT Press.
McLeod, D. and Adams, V. (1989). *Affect and Mathematical Problem Solving*. New York, NY: Springer.
Mellers, B., Hertwig, R., and Kahneman, D. (2001). Do frequency representations eliminate conjunction effects? An exercise in adversarial collaboration. *Psychological Science*, 12:269–75.

Montaño, U. (2014). *Explaining Beauty in Mathematics: An Aesthetic Theory of Mathematics*. Dordrecht, The Netherlands: Springer.

Montaño, U. (2012). Ugly mathematics: Why do mathematicians dislike computer-assisted proofs? *The Mathematical Intelligencer*, 34:21–28.

Netz, R. (1999). *The Shaping of Deduction in Greek Mathematics: A Study in Cognitive History*. Cambridge: Cambridge University Press.

Netz, R. (2005). The aesthetics of mathematics: A study. In P. Mancosu, K. F. Jørgensen and S. Pedersen (Eds.), *Visualization, Explanation and Reasoning Styles in Mathematics*, 251–93. Dordrecht, The Netherlands: Springer.

Netz, R. (2009). *Ludic Proof: Greek Mathematics and the Alexandrian Aesthetic*. Cambridge: Cambridge University Press.

Pascual, E. and Oakley, T. (2017). Fictive interaction. In B. Dancygier (Eds.), *Cambridge Handbook of Cognitive Linguistics*, 347–60. Cambridge: Cambridge University Press.

Paterson, G. (2013). *The Aesthetics of Mathematical Proofs*. Alberta: University of Alberta, MA thesis.

Pease, A., Lawrence, J., Budzynska, K., Corneli, J. and Reed, C. (2017). Lakatos-style collaborative mathematics through dialectical, structured and abstract argumentation. *Artificial Intelligence*, 246:181–219.

Poincaré, H. (1930). *Science and Method*. London: Thomas Nelson and Sons.

Raman-Sundström, M. (2012). Beauty as fit: An empirical study of mathematical proofs. *Proceedings of the British Society for Research into Learning Mathematics*, 32:156–60.

Rav, Y. (1999). Why do we prove theorems?. *Philosophia Mathematica*, 7:5–41.

Reber, R. (2016). *Critical Feeling: How to Use Feelings Strategically*. Cambridge: Cambridge University Press.

Reber, R., Brun, M. and Mitterndorfer, K. (2008). The use of heuristics in intuitive mathematical judgment. *Psychonomic Bulletin & Review*, 16:1174–78.

Reber, R., Schwarz, N. and Winkielman, P. (2004). Processing fluency and aesthetic pleasure: Is beauty in the perceiver's processing experience? *Personality and Social Psychology Review*, 8:364–82.

Reber, R. and Unkelbach, C. (2010). The epistemic status of processing fluency as source for judgments of truth. *Review of Philosophy and Psychology*, 1:563–81.

Reber, R., Winkielman, P. and Schwarz, N. (1998). Effects of perceptual fluency on affective judgments. *Psychological Science*, 9:45–48.

Rota, G. (1997). The phenomenology of mathematical beauty. *Synthese*, 111:171–82.

Sørensen, M. H. and Urzyczyn, P. (2006). *Lectures on the Curry-Howard Isomorphism*. New York, NY: Elsevier.

Schupbach, J. N. and Sprenger, J. (2011). The logic of explanatory power. *Philosophy of Science*, 78:105–27.

Silver, E. and Metzger, W. (1989). Aesthetic influences on expert mathematical problem solving. In D. McLeod and V. Adams (Eds.), *Affect and Mathematical Problem Solving*, 59–74. New York, NY: Springer.

Steiner, M. (1978). Mathematical explanation. *Philosophical Studies*, 34:135–51.

Todd, C. (2008). Unmasking the truth beneath the beauty: Why the supposed aesthetic judgements made in science may not be aesthetic at all. *International Studies in the Philosophy of Science*, 11:61–79.

Todd, C. (2018). Fitting feelings and elegant proofs: On the psychology of aesthetic evaluation in mathematics. *Philosophia Mathematica*, 1–19.

Topolinski, S. and Reber, R. (2010). Gaining insight into the "Aha" experience. *Current Directions in Psychological Science*, 19:402–05.

Wright, C. (1990). Wittgenstein on mathematical proof. *Royal Institute of Philosophy Supplement*, 28:79–99.

5

Can a Picture Prove a Theorem? Using Empirical Methods to Investigate Visual Proofs by Induction

Josephine Relaford-Doyle and Rafael Núñez

1 Introduction

Mathematics has been defined as "the science which draws necessary conclusions" (Peirce, 1881); to this end, formal proof is indispensable in modern mathematics. Proofs are the ultimate source of certainty in mathematics, and so the question of what may qualify as a valid proof is an important topic for philosophers and mathematicians alike. Today, most mathematicians would agree that a valid proof must take the form of a sequence of propositions in which each statement follows logically from the ones preceding. While visualization and images are widely appreciated as useful heuristic tools in mathematics, they were expelled from formal proof during the arithmetization movement of the late nineteenth century, and are not generally considered a valid means of mathematical justification. Today, the dominant view on the role of images in proofs is best captured by Moritz Pasch, who in 1882 explained that a theorem "is only truly demonstrated if the proof is completely independent of the figure" (Pasch, 1882/1926: 43).

However, in recent decades advances in visualization technologies have renewed interest in the use of visual representations in mathematics, including mathematical justification. Since 1976 the Mathematics Association of America has run a peer-reviewed "Proofs Without Words" column in their magazine, featuring specially designed images that are intended to demonstrate or "prove" a mathematical theorem. "Visual proofs"—images with minimal notation and no accompanying text—exist for theorems in a wide variety of mathematical domains including algebra, geometry, number theory, and calculus, and have been compiled into a number of volumes (Nelsen, 1993, 2000, 2016). While visual proofs are certainly items of interest within the mathematical community, most mathematicians and philosophers of mathematics would agree that they do not actually qualify as valid proofs. However, there is a small but vocal contingent that has argued for an expanded role of images in mathematical justification, with some going so far as to claim that "pictures can prove theorems" (Brown, 2008: 26).

The debate about the status of visual representations in mathematical justification has been waged almost entirely within the philosophy of mathematics and has received little attention from fields like cognitive science and psychology. However, many of the arguments made on both sides of the debate are actually empirical in nature—they are claims that can be explored, supported, or refuted using empirical methods. By observing the ways in which people use an image to justify a mathematical theorem, and by carefully examining the conclusions that people draw from a visual proof, it is possible to assess the extent to which an image may function as a proof and the conditions in which it is most likely to do so.

In this chapter we discuss findings from a study that was inspired by the philosophical debate over the status of images in mathematical proof, and which we designed to shed light on the ways in which people can use an image to justify a mathematical statement (Relaford-Doyle and Núñez, 2017). In the first part of the chapter we briefly summarize the main arguments that have been offered both against and in support of the use of images in mathematical proof. Then we turn to the specific case study of mathematical induction and "visual proofs by induction," which have received considerable attention within the debate and which we investigated in our study. We then describe a new method that we developed to investigate people's reasoning with a visual proof by induction. We discuss key findings from this study that we believe will be of interest to philosophers concerned with the use of images in mathematical justification, and discuss the relevance of these findings for the debate on the epistemic status of visual proofs in mathematics.

2 The debate over images

The debate surrounding the status of images in mathematical justification is long-standing and complex. Here we briefly outline the major arguments that have been made; for more extensive summaries, see Doyle et al. (2014) and Hanna and Sidoli (2007).

While visualization is an important and valued aspect of mathematical practice, images have not been accepted in formal proof since the arithmetization movement of the early twentieth century. What accounts for the demoted status of images in mathematical proof? Arguments against visual representations fall into three main categories:

- Images can be misleading or lead us to draw false conclusions. Images were expelled from formal proof in the late nineteenth century after several developments in formal mathematics seemed to be at odds with visual intuitions (for instance, the discovery of objects like space-filling curves and continuous but nowhere-differentiable functions). During this time, visualization and images came to be seen as unreliable and therefore unacceptable in mathematical proof (see Davis, 1993, and Mancosu, 2005, for a more detailed historical account). Today's math students are warned "not to trust the picture," and well-known examples of misleading images are presented as "cautionary tales" about the dangers of visual reasoning (see, for example, Giardino's 2010 discussion of Klein's "proof" that all triangles are isosceles).
- There is no standard method for "reading" an image, and so interpreting the argument that an image represents can be difficult and inconsistent between viewers. Unlike formal proofs, which make the chain of logical dependencies explicit by the order in which propositions are sequenced, images are presented "all at once" (Arcavi, 2003). Each viewer must determine for him- or herself which aspects of the picture are relevant to the argument. The potential for "wiggle room" in the interpretation of the image is clearly undesirable for a mathematical proof (Doyle et al., 2014). This has led some to suggest that propositional representations are not just the form of proof currently in favor within the mathematical community, but are actually more "ergonomic" forms of mathematical justification (Coppin and Hockema, 2009).
- Pictures are necessarily finite and can only display a single case or, at best, a small number of examples of a theorem. Examples can allow for inductive reasoning, in which we infer a general rule from particular examples, and can thus be a useful means of mathematical discovery. However, inductive reasoning is not acceptable in mathematical justification; you cannot demonstrate that a general theorem is true by pointing to particular examples. Images display examples pictorially, and while this can aid in understanding the theorem, the image itself cannot possibly act as a proof.

(This particular drawback of images has received extensive attention from the other side of the debate; for responses see Giaquinto, 2007 and Kulpa, 2009, who describe ways in which general conclusions *can* be reliably drawn from specific images).

For these reasons, most philosophers of mathematics today reject images as a valid means of proof. However, some are more sympathetic to visual representations and argue that images play an important role in mathematical justification. Rather than responding directly to the criticisms outlined above, these scholars argue that a strict adherence to formal, propositional proof as the only certification of mathematical truth is unrealistic. While not denying the importance of proofs in modern mathematics, this group points out that mathematical *practice*, including proof-writing and comprehension, is much wider and richer than just the production and reading of syntactic, propositional proofs. Philosophers concerned with mathematical practice have noted that the function of a proof is not just to verify the truth of a theorem, but to understand why the theorem holds, to "search for reasons" (Rota, 1997: 146), to "display the mathematical machinery" (Rav, 1999: 13) behind the result. Both propositional and visual reasoning can aid in this pursuit, so both should be acknowledged as playing a role in real-world mathematical justification (Giardino, 2010).

A smaller contingent has gone a step further, claiming that some images can actually stand alone as full-blown proofs. The primary argument offered in favor of this view is simple: in some cases, images are simply so clear, so convincing, that they allow the viewer to draw the correct mathematical conclusion "almost by inspection" (Davis, 1993: 336). Doyle et al. (2014) call this the "Romantic" perspective on proof: that visual proofs "can be far more rapidly and deeply convincing than traditional, propositional mathematical argumentation, and are therefore (in such cases) perfectly acceptable, even occasionally preferable proofs." James Robert Brown, the most vocal proponent of the idea that an image can prove theorems, claims that pictures can act as "windows into Plato's Heaven" allowing us to see with our mind's eye the truth of a mathematical theorem (2008: 40).

The debate over the epistemic status of images in mathematics often makes claims about general objects: can *a picture* prove *a theorem*? But, as mentioned earlier, there are hundreds of visual proofs that exist for theorems across many domains in mathematics. In order to begin to investigate any of the claims above empirically, we must "zoom in" our attention to a particular class of visual proofs. We selected "visual proofs by induction"—images designed to prove theorems that could be formally proven using mathematical induction—as a good starting point for empirical work. We made this selection for two reasons. First, visual proofs by induction are among the simplest and most elegant visual proofs and are often described as being particularly convincing (Brown, 2008; Chihara, 2004). Second, despite its ubiquity and importance in mathematics, formal mathematical induction has some notable drawbacks. It requires the use of complex algebra and nuanced notation, and is a notoriously difficult method for students, even educated adults, to learn (Avital and Libeskind, 1978; Fischbein and Engel, 1989; Movshovitz-Hadar, 1993). Additionally, a proof by mathematical induction is not always explanatory—while it can put the truth of a theorem beyond doubt, it may not provide insight into *why* the theorem is true (Lange, 2009; Hanna, 2000; Stylianides et al., 2016). Visual methods may be more accessible than formal mathematical induction, and may also provide deeper understanding into the truth of a result, and therefore present a promising starting point for empirical investigation. Below we present an example of a visual proof by induction and use it to dig deeper into some of the issues surrounding the use of visual representations in mathematical justification.

3 A visual proof by induction

Consider the following statement: "The sum of the first n odd numbers is equal to n^2." How might we go about deciding if this statement is true? We could start by testing some examples. The first six cases are given below:

$$1=1^2$$
$$1+3=4=2^2$$
$$1+3+5=9=3^2$$
$$1+3+5+7=16=4^2$$
$$1+3+5+7+9=25=5^2$$
$$1+3+5+7+9+11=36=6^2$$

In each case the statement holds true, and so we might conclude that the theorem is possibly or perhaps even probably true for all natural numbers n. However, these examples alone do not constitute a proof of the general theorem—we cannot conclude with certainty that the theorem is necessarily true for all natural numbers. Without a formal proof, our conclusion remains uncertain and open to the possibility of counterexamples.

This statement could be formally proven using mathematical induction, which despite its name is a deductive formal proof method that can be used to demonstrate that a theorem is necessarily true of all natural numbers. It does this in two steps: first it is shown by direct verification that the statement is true of a base case, generally $n = 1$. Then in the inductive step it is shown that if the theorem is true for some arbitrary natural number k, then it must also be true for its successor $k + 1$. This creates a sort of domino effect, extending over all natural numbers, which rules out the possibility of counterexamples and allows us to conclude that the general theorem is true for all n. A formal proof of our conjecture is given below:

Theorem: $1 + 3 + \ldots + (2n − 1) = n^2$
Base case: $n = 1 \rightarrow 1 = 1^2$
Inductive step: Assume $1 + 3 + \ldots + (2k − 1) = k^2$, for some fixed number k.
Adding the next odd number $2k + 1$ to both sides of the equation, we have:
$1 + 3 + \ldots + (2k − 1) + (2k + 1) = k^2 + (2k +1)$
Rewriting the last odd term and factoring the right side gives us:
$1 + 3 + \ldots + (2k − 1) + [2(k+1) − 1] = (k + 1)^2$, QED

Next, let's consider a "visual proof" of the same theorem, which is given in Figure 5.1 (drawn after Brown, 1997).

In the image, consecutive odd numbers of dots are arranged in layers, beginning with 1 in the lower left-hand corner. When the dots in the first n layers are considered together the resulting array forms an n by n square, and so the total number of dots in the array is given by n^2. While the image displays only the first six cases of the general theorem, a viewer might infer that the pattern will continue to hold as more layers are added, and therefore be convinced that the general theorem is true. Indeed, visual proofs by induction like the one in Figure 5.1 have been described as "completely convincing" (Chihara, 2004).

While a formal proof by induction requires complex notation and sophisticated algebra, visual proofs by induction can be quite simple—for instance, interpreting Figure 5.1 seems to require understanding only a few relatively simple concepts like odd numbers, addition, and squaring. The apparent simplicity of the visual proof has led some to suggest that the image can function as a proof even for nonexpert viewers who are unfamiliar with formal mathematical induction or mathematical proof in general: Jamnik (2001) claims that interpreting the image in Figure 5.1 as a proof requires only "basic secondary school math

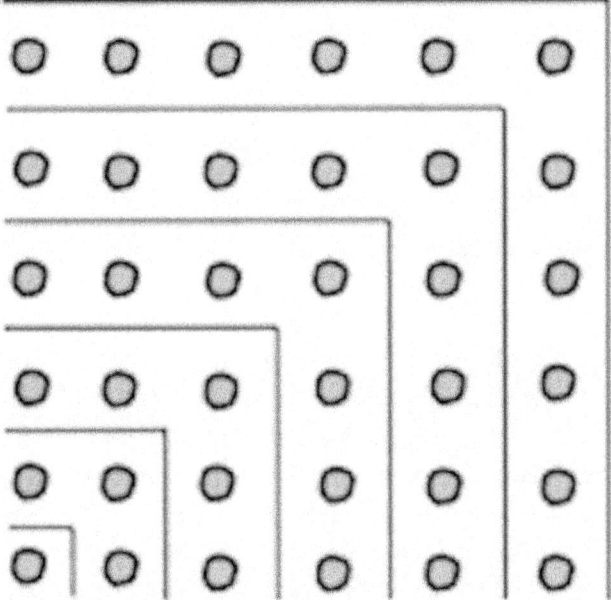

Figure 5.1 Visual proof that $1 + 3 + 5 + \ldots + (2n - 1) = n^2$ (drawn after Brown, 1997).

knowledge," while Brown (2008) cites anecdotal evidence that "people with no particular training in mathematical induction grasp the picture proof with no difficulty" (43). The claim that visual proofs by induction can provide proof-like experiences even for nonexpert viewers is a particularly interesting one, as it seems to suggest that there could be ways to achieve mathematical certainty, *outside* of formal mathematics.

Visual proofs by induction have often been described as "convincing," but the exact nature of the conclusions that viewers draw from the image has not been investigated. It may be hasty to assume that the image provides viewers with the same degree of certainty associated with formal proof. After all, the image only shows a finite number of cases of the general theorem; for instance, the visual proof in Figure 5.1 shows that the sum of the first n odd numbers is equal to n^2 only up to the $n = 6$ case. This might suggest that the image in Figure 5.1 would serve a similar function as the six numerical examples presented earlier—it may convince the viewer that the theorem *likely* holds for all natural numbers, but cannot provide certainty. On the other hand, the image contains structure that is not available in the same cases presented numerically, and which could be exploited in order to demonstrate that the property would *necessarily* continue to hold for values not depicted in the image. Specifically, it can be demonstrated

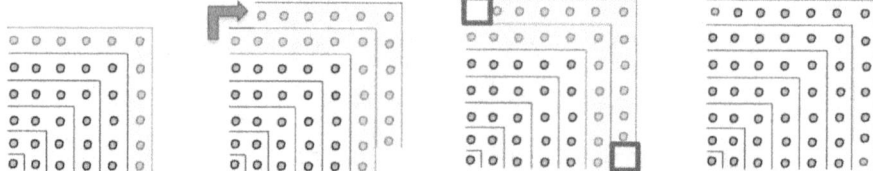

Figure 5.2 Rigorous image-based justification of the general theorem, showing that the next layer must contain the next consecutive odd number of dots to maintain the square shape.

that the square shape is preserved if and only if the next layer contains the next consecutive odd number of dots. There are various ways of doing this, but one is as follows: if we start with an $n \times n$ square, the $(n + 1)^{st}$ layer could be constructed by copying the nth layer and translating the copy up one unit and right one unit (Figure 5.2a,b). This results in two vacant positions that must be filled in order to maintain the square shape (Figure 5.2c,d). Thus, every layer must contain exactly two more dots than its predecessor. Since the difference between any two consecutive odd numbers is 2, we can conclude that the new layer must contain the next consecutive odd number of dots.

Though clearly not a traditional propositional proof, it could be argued that an argument such as this does establish the truth of the general theorem for all natural numbers, and that this conclusion meets the level of certainty required for mathematical proof. To distinguish this argument from a formal proof, we will refer to it (and any equivalent argument) as a "rigorous image-based justification" of the theorem.

It is unknown how accessible such arguments are to viewers, and, more generally, how closely the conclusions drawn from the image resemble the conclusions drawn from a formal proof. In the study discussed below we applied empirical methods to assess the extent to which a visual proof by induction may serve a proof-like function for viewers. Specifically, we asked three key questions: (1) To what extent can the image stand alone as a proof of the theorem? (2) Is the image "convincing"? What is the nature of conclusions drawn from the image, and to what extent do those conclusions resemble those from formal proof? (3) Can viewers construct rigorous image-based justifications for the general theorem, and if so, do they consider these arguments valid? Finally, we also explored whether expertise—specifically, familiarity with the formal proof method of mathematical induction—has an impact on any of these outcomes, or if the image really is accessible even to viewers who have not been trained in formal proof.

4 Method

We carefully devised an empirical study to address the three questions outlined above. Our study (Relaford-Doyle and Núñez, 2017) consisted of an open-ended task, followed by a structured interview, both of which were designed to explore many of the empirical claims that have been made in the philosophical debate over the status of visual proofs.

As mentioned above, some have claimed that some visual proofs by induction are accessible and convincing even to viewers without any particular training in mathematics, and specifically to viewers who are not familiar with formal mathematical induction (Jamnik, 2001; Brown, 2008). We found this to be a particularly provocative claim, and one that could be tested empirically by comparing the reasoning of trained and untrained viewers with the visual proof. With this in mind, we recruited participants from two distinct populations. Our first group of participants was drawn from the general subject pool and consisted of university-level students with a variety of majors including psychology, cognitive science, and linguistics. None of these students had taken a university-level mathematics course on proof-writing, and so we call these participants "proof-untrained." Importantly, despite not having taken a course on proofs, these participants were all highly educated adults enrolled at a prestigious university, such that we expected them to have a firm grasp of the relatively simple mathematical concepts involved in the statement (like odd numbers, addition, and squaring). We recruited our second group of participants through the mathematics department—specifically, we enrolled individuals who had received at least a B- in "Mathematical Reasoning," an upper-division mathematics course that covers a variety of formal proof methods including mathematical induction. We refer to this group of participants, mostly mathematics majors, as "proof-trained." In order to capture what we expected would be a wide variety of responses to the task, we recruited a relatively large number of participants in each group—twenty-five proof-untrained participants, and twenty-four proof-trained.

We were also interested in exploring how much information needs to accompany the image in order for it to function as a proof. It has been argued (Giardino, 2010) that a "Proof Without Words" is not really "without words" at all, since it likely needs to be accompanied at least by the statement it is intended to prove. To determine to what extent the image can "stand alone" as a proof, we devised three distinct conditions. In all three conditions we gave our participants a task that included the visual proof in Figure 5.1, but we varied the amount of information that participants had about the image. In our Justification condition, we provided the full target statement ("The sum of the first n odd numbers is equal to n^2") and asked the participant to explain how the image showed that the statement was true. In our Supported Discovery condition the participant was given a fill-in-the-blank version of the target statement ("The sum of the first n odd numbers is equal to _____.") and asked to fill in the blank and explain how they got their answer. In the Full Discovery condition, participants were told that a mathematician drew the picture while trying to prove a statement about the sum of odd numbers. We asked them to guess what the mathematician was trying to prove and explain their answer. These three conditions also allowed us to explore whether our participants, particularly our proof-trained students, would be sensitive to the differences between contexts of mathematical discovery (in which the use of images and inductive reasoning is acceptable) and justification (in which it is not).

We gave each participant a worksheet with their instructions and the visual proof, and explained that their ultimate goal was to create a "tutorial video" in which they explained their response to the task to an imagined third-party audience. We asked them to imagine that someone else would watch their video, and to be as complete and helpful to that person as possible. Prior to filming their tutorial video, participants could take as much time as they needed to think about the task and plan their response. During this time they had access to pens, pencils, highlighters, and additional paper, and were free to mark the worksheet in any way that they felt helpful. Once they were ready the participant explained their response to the task in their tutorial video. Both the planning and explanation stages were entirely self-paced and occurred without the researcher present. The entire process was recorded by a camera mounted directly above the participant's workspace, giving us access to their speech, writing, and gestures toward the worksheet (Figure 5.3).

Being a relatively open-ended task, it is no surprise that the tutorial videos our participants produced provided an extremely varied and rich source of data. In order to address our specific research questions, two independent coders rated each tutorial video on three dimensions:

1. *Understanding of target statement.* We assessed whether the participant had demonstrated understanding of the statement that the image was intended to prove (i.e., that the sum of the first n odd numbers is equal to n^2).
2. *Use of image as justification.* We examined the strategies that the participant used to explain how the image showed that the statement

Can a Picture Prove a Theorem?   105

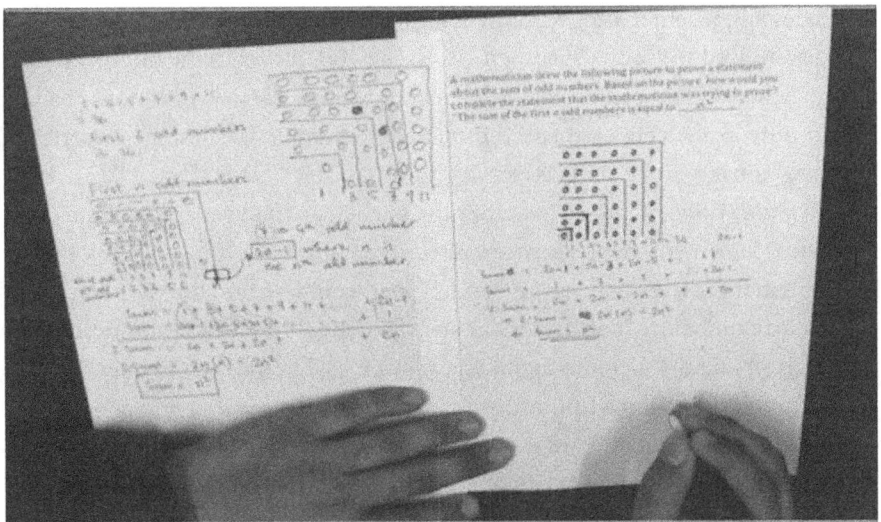

Figure 5.3 Still image from video data showing a participant's workspace.

was true. We distinguished between case-based strategies, in which the participant used the image to demonstrate one or multiple particular cases of the theorem, and pattern-based strategies, in which the participant described a general pattern in the image which could hold beyond the first six cases that the image displayed. We considered case-based arguments to be indicative of inductive reasoning, while pattern-based arguments more closely resembled formal proof in that they at least suggested generality of the theorem. We also noted participants who provided a rigorous image-based justification like the one described in Figure 5.2.

3. *Relevant features of image.* We kept track of which features of the image the participant mentioned in their tutorial video, including the consecutive odd numbers of dots in each layer, the square shape of the image, and the idea that the image shows the first cases of a pattern that could be extended.

The tutorial videos provided us with information about the variety of ways in which viewers used the visual proof to justify a mathematical statement. The second part of our study consisted of a more structured interview that the researcher conducted with each participant after they had filmed their video. The purpose of the interview was to assess in a standardized way the extent to which the image had served a proof-like function for each participant. Specifically we were interested in assessing whether after working with the visual proof the participant had generalized the target statement to cases not depicted in the image, and if so, whether that conclusion was considered necessary, as in formal proof, or only likely, as in case-based inductive reasoning. We designed a series of questions to probe these issues:

4. *Generalization.* To assess whether the participant had generalized the target statement, the researcher asked two questions: "Do you think the statement is true in all cases?" and "What would be the sum of the first 8 odd numbers?" If a participant answered "yes" to the first question and applied the formula to quickly answer "64" to the second question (without doing the addition), we characterized them as having generalized the target statement. Participants who answered "no," "I don't know," or used addition to calculate the sum of the first 8 odd numbers were characterized as not having generalized.

5. *Necessity.* Next, we needed to assess whether this generalization carried with it the level of necessity associated with formal mathematical proof. Importantly, we didn't want to assume that simply being willing to say the statement was true "in all cases" implied mathematical certainty. The word "all" is used quite freely in daily life. For instance, if we hear someone say "All Californians love the beach," it is very unlikely that the speaker actually means that every single Californian without exception enjoys the beach. Instead, "all" in such cases is used to mean "most" or "the majority." Obviously, this is not the same meaning as the universal quantifier "all" in mathematics, which implies the impossibility of counterexamples. Thus, to probe the sense in which our participants believed the statement to be true "in all cases," we created a follow-up question for any participant who had indicated generalization: we raised the possibility of large-magnitude counterexamples—"very large numbers where the statement actually isn't true"—and asked the participant what they thought about this suggestion. We gauged each participant's resistance to this suggestion on a 0–5 scale. A score of 0 indicated that the participant showed no doubt that counterexamples to the theorem could exist, and was therefore associated with low mathematical necessity. A score of 5 was reserved for participants who stated counterexamples were impossible, thus demonstrating a high degree of mathematical necessity.

Finally, any participant who showed a high degree of doubt regarding the possibility of counterexamples was asked how they might argue against such a claim. We examined the extent to which their arguments relied on the image, and noted any participants who produced rigorous image-based justifications comparable to the one given in Figure 5.2.

5 Results

5.1 Can the image stand alone as a proof?

In our three task conditions we provided participants with different amounts of information about the mathematical statement that the image was designed to prove. When the full theorem was provided (our Justification task), we observed that all participants in both groups demonstrated understanding of the statement. However, when less information was provided, we saw proof-untrained participants were significantly less likely to generate the target statement than proof-trained participants. Figure 5.4 shows the proportion of participants from each group that demonstrated understanding of the target statement across our three conditions:

In our Supported Discovery condition, all proof-trained participants were able to determine that the sum of the first n odd numbers is n^2; 6/8 (75 percent) proof-untrained participants correctly completed the statement. Performance dropped for both groups in our Full Discovery condition: 6/8 (75 percent) proof-trained participants were able to generate the full target statement, while only 3/9 (33 percent) proof-untrained participants arrived at the target statement. Thus, while proof-trained participants were likely to understand the

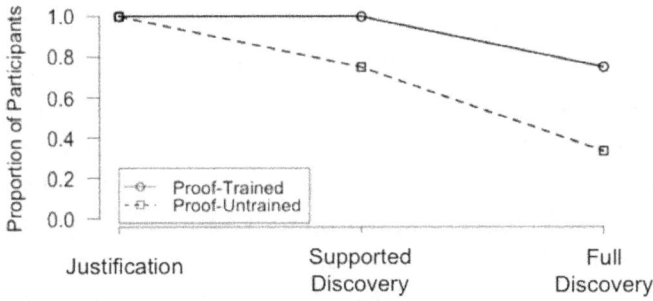

Figure 5.4 Proportion of participants who demonstrated understanding of the target statement in each of the task conditions. Reprinted from Relaford-Doyle and Núñez (2017).

mathematical statement that the image is intended to prove even when it was not provided, proof-untrained participants seemed to need significant support to connect the image to the intended mathematical theorem. For most untrained viewers the image must be accompanied by the theorem it is intended to prove, or at least a substantial hint, in order for it to even potentially function as a proof of that theorem.

If the proof-untrained participants didn't discover the full target statement, what did they think the image was intended to prove? Most of the proof-untrained participants in the Full Discovery condition concluded that the image had been drawn to demonstrate a fact about odd numbers: either that odd numbers increase by 2 (two participants), or that the sum of two odd numbers is an even number (three participants). While these are both true mathematical statements, they are notable in that they leave out any mention of n^2, and so fail to make use of the square shape of the array; it is as if the square-ness of the image was irrelevant for these participants.

Surprisingly, we found that this lack of attention to the square shape was not unique to the Full Discovery condition; across all conditions, of the seventeen proof-untrained participants who demonstrated understanding of the full target statement, only seven (41 percent) stated in their tutorial video that the square shape was a relevant feature of the image. In other words, even when participants knew that the picture is intended to represent that the sum of the first n odd numbers is n^2, the majority never mentioned that the dots in the image form a square. Instead, most proof-untrained participants explained that the number of dots in the array could be calculated by multiplying row × column (which is not unique to a square array), or simply manually counted dots to verify specific cases of the theorem. The fact that at each step the layers of the image are arranged to form a square shape—an observation that is essential for interpreting the image as a proof—went unmentioned by a surprising number of proof-untrained participants (in contrast, 95 percent of the proof-trained participants who correctly accessed the target statement mentioned the square shape during their tutorial video).

One possibility is that these proof-untrained participants *did* notice the square shape, but for whatever reason didn't describe it in their explanation—we cannot conclude that they didn't *see* the square just because they didn't *mention* it. However, further analysis of participant work suggests that in many cases the square shape truly went overlooked. Proof-untrained participants who redrew the image during their explanations often created drawings that violated the row-column structure array and the resulting square shape.

Figure 5.5 Work of two proof-untrained participants who redrew the image in a way that violated the row-column structure and square shape of the array.

Figure 5.5 shows drawings from two such participants. These drawings clearly do not maintain the regular grid structure of the original image; while they still contain thirty-six dots, the rows and columns have broken down and the dots are now arranged in an irregular pattern. The willingness to violate the grid structure in their explanatory drawings indicates that these participants likely did not notice the square shape as an important feature of the visual proof.

We were surprised that so many of our proof-untrained participants—highly educated adults with extensive experience in the basic arithmetic concepts involved in this theorem—didn't see that the square shape was an essential feature of the image even when they knew it was intended to represent a theorem involving n^2. Despite their high level of education, these viewers may need even more guidance to fully understand the image—for instance, they may need someone to point out the square shape and explain its connection to the theorem. In any case, it is clear that the image is surprisingly difficult for most untrained viewers to interpret, and is far from standing alone as a visual proof of the theorem for these viewers.

5.2 What conclusions do viewers draw from the image, and do these conclusions carry the necessity associated with formal proof?

Those who argue that pictures can act as proofs routinely point to how pictures can be deeply "convincing"—yet little is known about the conclusions that viewers actually draw from the image and the extent to which these conclusions resemble those of a formal proof. Does the visual proof convince viewers that the theorem is true *for all n*?

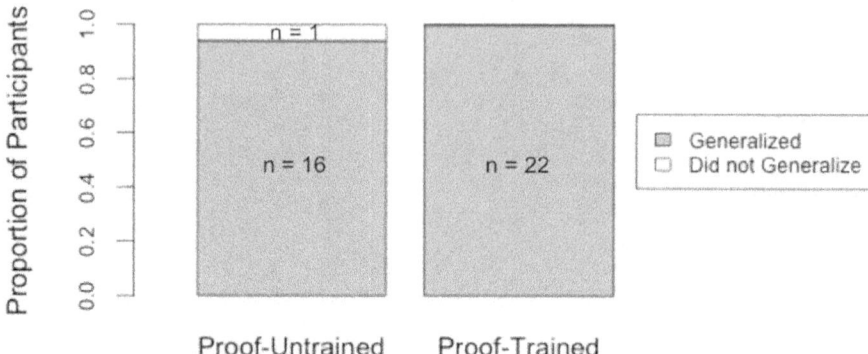

Figure 5.6 Participants in both groups were willing to generalize the target statement to nearby cases like $n = 8$. Reprinted from Relaford-Doyle and Núñez (2017).

Clearly, the image can only be convincing to a viewer who has identified the theorem that the image is intended to prove. Therefore, in this analysis we focus on the seventeen proof-untrained participants and twenty-two proof-trained participants who either were given or successfully generated the target statement. In the interview portion of the study, we asked each of these participants if they believed the statement was true "in all cases," and what the sum of the first 8 odd numbers would be. Based on these two questions, virtually all participants—both proof-trained and untrained—indicated willingness to generalize the theorem to cases not depicted in the image (Figure 5.6).

So, regardless of familiarity with formal proof methods, participants *did* say that they believed the theorem to be true "in all cases," and were willing to apply the rule to a case that was not depicted in the image. Can we conclude that the image functioned as a proof of the general theorem for these viewers? Caution is necessary here; as discussed earlier, the word "all" is polysemous; it is used in daily life to mean "most," "the majority," or "in general," none of which are synonymous with the meaning of "all" as the universal quantifier in mathematics. Similarly, the $n = 8$ case is very close to the cases provided; should we assume that that these viewers would be equally certain that the rule applies to $n = 300$, 10327, or 10^{26}?

To dig deeper into the nature of the conclusions our viewers had drawn, we next suggested the possibility of large-magnitude counterexamples: "Very large numbers where the statement actually isn't true." Here our results revealed a statistically significant difference between the two groups, with proof-trained participants being significantly more likely to indicate high resistance to the suggestion of counterexamples (Figure 5.7). Out of the sixteen proof-untrained

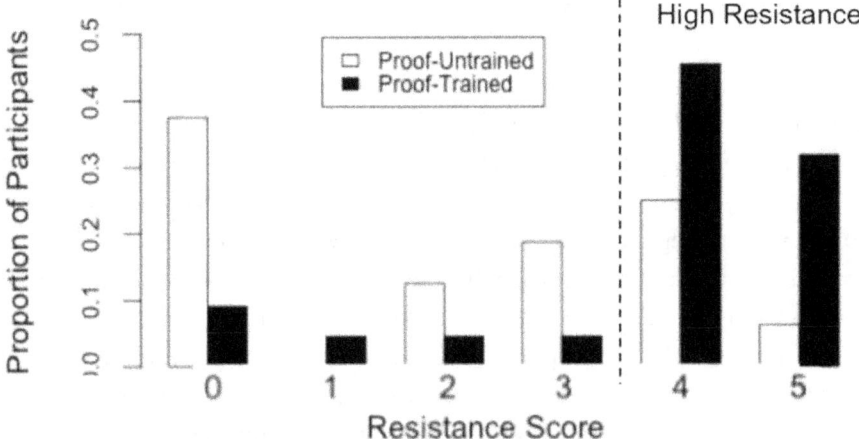

Figure 5.7 Proof-trained participants were significantly more likely to show high resistance to the suggestion of large-magnitude counterexamples than were proof-untrained participants. Adapted from Relaford-Doyle and Núñez (2017).

participants who generalized the theorem, seven (almost 40 percent) expressed no doubt whatsoever that there could be large-magnitude counterexamples to the theorem. Only five (31 percent) proof-untrained participants showed high resistance to counterexamples (characterized as a score of 4 or higher). In sharp contrast, seventeen out of twenty-two (77 percent) proof-trained participants were highly resistant to the possibility of large-magnitude counterexamples to the theorem.

This effect becomes even more striking when we consider the fact that some of our proof-untrained participants, despite not having taken the math reasoning course, were nevertheless familiar with mathematical induction. We had three participants in the proof-untrained group who indicated familiarity with mathematical induction on an exit survey that they completed following the interview portion of the study. When we look at these participants' resistance to counterexamples, we see that they resembled the proof-trained group—all three showed high resistance. In sum, we had only two participants who were genuinely unfamiliar with formal mathematical induction who showed high resistance to counterexamples after working with the visual proof.

These results suggest that for most proof-untrained participants the image serves as the basis of a standard inductive generalization; it displays examples from which the viewer generalizes a rule, but does not provide certainty or convince the viewer of the necessity of the theorem. The conclusion that proof-untrained participants utilize the image as a set of examples is supported by the finding that most proof-untrained participants used case-based strategies in their explanations—rather than describing a general pattern, they used the image to explain one or two particular cases of the theorem. Only proof-trained participants are likely to generalize the theorem *and* resist the possibility of counterexamples. These empirical findings—obtained with highly educated individuals—refute claims by Brown (2008) that the image is equally accessible to all viewers regardless of their familiarity with mathematical induction, as well as Jamnik's (2001) assumption that interpreting the image as a proof requires only "basic secondary school knowledge of mathematics." Finally, this result highlights the importance of careful empirical work; while virtually all participants *appeared* to have generalized the target theorem to "all cases," deeper investigation revealed that "all" meant something very different to the two groups.

5.3 Can viewers use the image to rigorously justify the general theorem?

So far we have shown that, after working with the image, a small minority of proof-untrained viewers and a larger proportion of proof-trained participants concluded that the theorem is true for all n. This suggests that the image *may* be serving a proof-like function for these viewers—however, in order to truly assess this we need to look at ways in which participants used the image to justify the truth of the general theorem. Can participants produce an argument like the one in Figure 5.2, to demonstrate that the pattern would *necessarily* continue to hold?

Very few participants produced such arguments, with proof-trained participants being more likely to do so. No proof-untrained participants spontaneously provided a rigorous image-based justification of pattern extension in their tutorial video. During the interview only two proof-untrained participants, SL and JY, were able to explain why the next layer of the image must always contain the next consecutive odd number of dots. SL was one of the participants described earlier who, despite not having taken the mathematical reasoning course, was familiar with mathematical induction. When first presented with the suggestion of large-magnitude counterexamples SL responded by writing a formal proof by mathematical induction, and only produced a rigorous image-based justification after many minutes of thinking and significant prompting by the interviewer. JY, on the other hand, responded to the suggestion of counterexamples by quickly producing the same rigorous image-based argument as was given in Figure 5.2. The fact that JY was able to do this, despite being genuinely unfamiliar with formal mathematical induction, shows that knowledge of the formal method is not prerequisite to being able to use the image to produce a rigorous argument.

Proof-trained participants were more likely than proof-untrained participants to produce rigorous justifications using the image, but it was still only a small number of participants who were able to do so. Five proof-trained participants provided a rigorous image-based argument as part of their tutorial video, and during the interview three more were able to explain why the pattern in the image necessarily continues. So, while sixteen proof-trained participants were highly resistant to the possibility of counterexamples, only half used the image to explain why counterexamples are impossible. A larger proportion referenced formal mathematical induction to justify the truth of the general theorem—75 percent of proof-trained participants mentioned that they could prove the general theorem using mathematical induction, and 25 percent actually wrote a formal proof, even though that was not a requirement of the task.

Proof-trained participants, including those who produced rigorous image-based arguments, expressed a variety of opinions about whether the image was a valid justification of the general theorem. Only one participant, RB, unhesitatingly indicated that the image could function as a proof of the theorem. In his tutorial video RB produced a particularly elegant argument in which he used the image, rather than symbolic notation, to perform a proof by induction. He begins the video by stating, "This problem is a pretty standard mathematical induction problem. So to do that, we're gonna start with the base case.... So the first case would be that n is 1. The first square made up of circles is 1 squared, which is 1." As he speaks, he draws a single circle, underlining it to highlight it as the first case of the theorem (Figure 5.8). He also verifies the base case in symbolic notation.

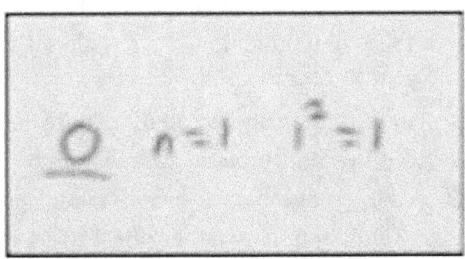

Figure 5.8 RB demonstrated the base case both visually and symbolically.

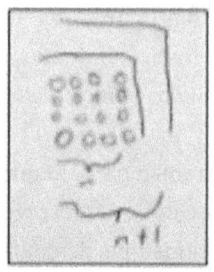 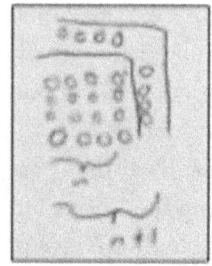 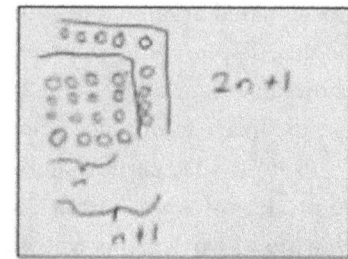

Figure 5.9 RB used the image to perform the inductive step.

Next RB performs the inductive step: "So let's assume it's true for some n. So you have some n by n square of circles made up like this." At this point RB draws a 4-by-4 square, but explains "I'm using 4 as an example, but this works with any n." He then describes a process for constructing the next layer: "We want to add a new outer layer, basically. Like the drawing shows, we're going to add a new layer and make it go from n to $n + 1$. (Figure 5.9a) So the question is, how many circles are we adding?" RB then shows why the next layer must contain the next odd number of dots: "We need to add at least n on this side (*gesturing down along the new column, copying the 4 dots from the previous square*) and n on this side (*gesturing across the new row on top, again copying the four dots, Figure 5.9b*). So we have n, n, and one more to complete the corner (*adds the corner circle, Figure 5.9c*). So the number we added was $2n + 1$ .... So we're adding a new odd number, the $(n + 1)^{th}$ odd number to it. By adding the new odd number we're getting to the new $(n + 1)$ by $(n + 1)$ square, which is just $(n + 1)^2$."

Later when RB is asked about the possibility of large-magnitude counterexamples he shows a high degree of resistance, saying, "I don't see how that would happen, cause the inductive proof would make it true for all n." Interestingly, as he says "inductive proof" he points to the drawing he's produced, suggesting that he considers his image-based argument to be a full-fledged proof by induction. He later explains, "The picture is actually a pretty good representation of the actual induction , . . . It makes it a lot easier to picture how you go from n to $n + 1$."

RB spontaneously used the image to demonstrate a rigorous inductive argument, and considered this image-based approach to be basically equivalent to the formal proof. A larger number of proof-trained participants ($n = 6$), however, expressed tension around whether the image could function as a proof. Like RB, these participants were able to use the image to explain why the next layer had to have the next odd number of dots, but unlike RB they didn't immediately recognize this argument as rigorous and expressed a clear preference for the formal proof. For instance, SN initially justified her belief that the theorem was true "in all cases" by referring to the image: "I can know for sure that the next segment will contain the next odd number . . . this picture will be applicable for all n values." However, when the interviewer raises the possibility of counterexamples, SN turned to the formal method, saying, "There's this other proof that we do in our classes, proof by induction. I haven't done that for this, but I could try." She then completed a formal proof by induction, only after which she concluded that counterexamples are impossible.

By referring to proof by induction as "this *other* proof," SN seemed to suggest that she considered the argument she has already provided—an image-based justification—to be a proof, but one that is distinct from mathematical induction. On the other hand, SN wasn't willing to fully reject the possibility of counterexamples until she had produced the formal proof. There's an obvious conflict here: if the image is a proof, then the proof by mathematical induction should be completely redundant and unnecessary—why did she feel the need to perform the formal proof? It seems as if, while she recognized the image as proof-like, SN clearly did not hold it in the same regard as formal proof. Other proof-trained participants expressed similar tension:

- When asked if his image-based argument proved the theorem, JC answered "Not rigorously, but pictorially." He later explained that, "In one sense, you can [use the image to prove the theorem], because this is a visual version of rigorous proof. But in another sense you can't, because you can't do it concretely."
- In his tutorial video, RG first produced a rigorous image-based justification for the general theorem, then produced a formal proof using Gauss's Theorem to "further prove" the theorem. After completing the formal proof, he remarked that the image actually had already provided "mathematical proof," but "using Gauss's Theorem is more of a formal mathematical proof using an already established theorem." He laughingly concluded that, once he had provided the image-based argument, the formal proof "isn't even really necessary."
- SY used the visual proof to provide a rigorous image-based justification of the theorem, but contended that this method "is not like a normal thinking way," and said she prefers formal proof "because you can think clearly, what you want to express."

Another group of participants ($n = 6$) completely rejected the image as a means of proving the theorem. These participants required a formal proof, arguing that because the image is finite it can't be used to show that counterexamples to the theorem are impossible. YZ explained, "The picture is only good for finite cases, you cannot draw the picture infinite times." XX argued, "By drawing patterns you cannot conclude anything, you cannot conclude that everything is true. The base case is true, but if n is really large you cannot draw a picture to show it's true." EL explained his rejection of the image: "I don't think I can use this picture [to argue against counterexamples]. A picture like this is for younger people, like for a college student to explain to a primary student . . . . But for the college student or the professor, we need to do that abstractly, use abstract concepts to explain the formula because the larger numbers we cannot represent by this picture." Interestingly, one participant rejected the visual proof even after producing a rigorous image-based justification, explaining, "You can't prove anything from just looking . . . I don't like to prove it using the picture. It can help with explaining. I guess I would use this picture in conjunction with 'mathier' things, not just the picture."

Thus, while proof-trained participants are more likely than proof-untrained participants to produce rigorous image-based arguments, it is still only a relatively small proportion that does so. Participants who produce such arguments generally do not immediately consider the image-based justification as a valid proof, and there is still a strong preference among most proof-trained participants for a formal proof by mathematical induction.

5.4 Summary of findings

In our study we asked three questions: (1) Can the image stand alone as a proof?, (2) What conclusions do viewers draw from the image, and do these conclusions carry the necessity associated with formal proof?, and (3) Can viewers use the image to rigorously justify the general theorem? For all three questions, the answer seems to depend on *who* is doing the viewing. Participants—well-educated college students—who were not familiar with formal mathematical induction often overlooked key features of the visual proof and would likely require additional support to understand the image. For proof-trained participants, the image came closer to "standing alone" as a proof, as the majority of these participants were able to identify the target theorem even when it was not provided. When participants did recognize the target theorem, our findings reveal that the proof-trained and proof-untrained participants drew different conclusions from the image. While participants in both groups agreed that the theorem was true "in all cases," proof-trained participants showed significantly higher resistance to the possibility of large-magnitude counterexamples. For proof-untrained participants the image seemed to function as a set of examples that allowed for a standard inductive generalization. Almost no proof-untrained participants produced rigorous image-based justifications for the general theorem. Proof-trained participants were more likely to produce such arguments, but only one participant unhesitatingly recognized the image-based justification as a fully valid proof. Most proof-trained participants expressed a preference for the formal proof strategy, and many explicitly rejected the image as a means of justifying the general theorem.

6 Discussion and final thoughts

The debate over the status of visual proofs in mathematics has been conducted almost exclusively within philosophy, but at the same time has produced a number of claims that could be tested empirically about the nature of reasoning with such images. In the study described above we applied empirical methods to investigate some of the claims that have been made about the use of visual representations in mathematical justification, focusing on the particular case of a visual proof by induction. While this study only begins to scratch the surface of reasoning with visual proofs, we believe these findings to be valuable to any philosopher who is concerned with the epistemic status of images in mathematics. At the same time, our results reinforce the crucial importance of informing philosophical inquiry with empirical work.

Our findings clearly refute claims that visual proofs by induction are equally accessible to all viewers, regardless of their familiarity with formal mathematical induction (Brown, 2008; Jamnik, 2001). For the majority of our proof-untrained participants, the image seemed to function as a set of examples; it allowed for an inductive generalization, but did not provide the degree of certainty associated with formal proof. This calls into question the argument that such visual proofs are immediately convincing *in a proof-like way*, even to nonexpert viewers. At the same time, observations of our proof-trained group revealed that viewers—particularly those familiar with formal methods—*can* use the structure provided in the image to produce rigorous image-based justifications for the general theorem. Rather than allowing for some direct, immediate perception of the truth of the general theorem, it seems that the image is most likely to be mathematically convincing when it can be integrated with propositional, formal knowledge—or, as Giardino says, when there can be a "continuous 'dialogue' between language and figures" (2010: 37). This was best demonstrated by our proof-trained participant RB, who spontaneously performed a proof by mathematical induction using the structure provided by the image, rather than algebraic notation.

The majority of our proof-trained participants were hesitant to treat the image as a proof. Notably, many participants exhibited reluctance, or even rejected the image flat-out, even when they had used it to provide a rigorous justification of the general theorem. We believe our proof-trained participants' simultaneous ability to generate rigorous image-based arguments *and* reluctance to accept these arguments is reflective of their training, which has undoubtedly favored the development of propositional notation-based methods. In regard to the larger question of whether pictures can prove theorems, we are inclined to agree with Kulpa's (2009) argument that images' perceived lack of rigor is due in large part to the simple fact that the mathematical community has focused almost exclusively on developing and refining methods of *propositional* justification, while devoting virtually no attention to developing reliable visual reasoning practices. Visual representations differ from propositional ones in several fundamental ways. For instance, algebraic arguments use variables to achieve generality; visual representations, however, have no built-in means of representing a general argument that could be applied to any case. Our proof-trained participant RB overcame this by including a crucial caveat in his verbal explanation of the visual proof: that, although his drawing used the $n = 4$ case, his argument could be applied to any n. It is likely that other proof-trained participants' preference for formal, algebraic proof stems from the lack of a clear means of generalization in the image—without a clear set of reasoning practices, most viewers remain wary of image-based justifications. However, RB's explanation demonstrates that visual proofs by induction provide a case in which visual reasoning methods could be "tamed" (were we so inclined).

Not surprisingly, the findings from this initial investigation raise a variety of additional questions. For instance, it is not immediately clear *why* the proof-untrained participants expressed less resistance to the existence of counterexamples than their proof-trained counterparts. It is possible that, without training in formal mathematical induction, the image is simply viewed as a set of examples and allows only for standard inductive generalization. However, it could also be the case that our proof-trained participants didn't fully understand the visual proof, and it was this that explains their response to counterexamples.

Moreover, our proof-untrained participants were likely unfamiliar with the general norms involved in mathematical justification (e.g., the meaning of the word "all" when used as the universal quantifier in a mathematical context, and the fact that proofs allow us to draw general conclusions with certainty). If this is the case, then we would expect that training—either in how to interpret the image, or in the general principles of mathematical proof, or both—would allow participants to draw proof-like conclusions from the image, while still being unfamiliar with the specific proof method of mathematical induction.

Additionally, in this study we investigated how university-level students used a visual proof to justify a mathematical statement. While this allowed us to assess claims about the role of expertise in interpreting a visual proof by induction, it leaves open many questions about how professional mathematicians would reason with and assess these images as proofs. A variety of studies have shown that university-level mathematics students have nonnormative views about proof (e.g., Harel and Sowder, 1998; Selden and Selden, 2003; Weber and Mejía-Ramos, 2014), and so we might expect that expert mathematicians would show a notably different pattern of responses than our proof-trained participants. On the other hand, other studies have reported similar patterns of results between mathematics professors and undergraduates in regard to the admissibility of graphical inferences in proofs (Zhen et al., 2016; Weber and Mejía-Ramos, Chapter 6), so perhaps this is a task in which mathematicians and math students would behave similarly. In any case, further work would clearly be needed to characterize how professional mathematicians interpret a visual proof by induction, whether they can use it to construct a rigorous justification, and what views they have about its validity as a proof.

Finally, our results highlight the value of informing philosophical arguments with careful empirical work. Philosophers of mathematics who are interested in visual proofs are presumably all themselves familiar with formal proof methods like mathematical induction—and, as our study revealed, their expertise fundamentally changes the nature of their experience with the image and the claims they can make based on primarily introspective methods. Thus, even for the simplest of visual proofs, a philosopher's experience with the image is likely very different than that of most people, including other highly educated adults. In determining whether pictures can prove theorems, we must be careful not to turn exclusively to our own experience for the answer, but instead seek out empirical data that can shed light on crucial questions: *Which picture? With what information? And for whom?*

References

Arcavi, A. (2003). The role of visual representations in the learning of mathematics. *Educational Studies in Mathematics*, 52(3):215–41.

Avital, S. and Libeskind, S. (1978). Mathematical induction in the classroom: Didactical and mathematical issues. *Educational Studies in Mathematics*, 9(4):429–38.

Brown, J. R. (1997). Proofs and pictures. *The British Journal for the Philosophy of Science*, 48(2):161–80.

Brown, J. R. (2008). *Philosophy of Mathematics: A Contemporary Introduction to the World of Proofs and Pictures*, second edition. New York, NY: Routledge.

Chihara, C. S. (2004). *A Structural Account of Mathematics*. New York, NY: Oxford University Press.

Coppin, P. and Hockema, S. (2009). A cognitive exploration of the "non-visual" nature of geometric proofs. In P. Cox, A. Fish and J. Howse (Eds.), *Proceedings of the Second International Workshop on Visual Languages and Logic*, 81–95, Corvallis: IEEE Computer Society.

Davis, P. J. (1993). Visual theorems. *Educational Studies in Mathematics*, 24(4):333–44.

Doyle, T., Kutler, L., Miller, R. and Schueller, A. (2014). Proofs without words and beyond. *Convergence*. Retrieved July 23, 2017 from http://www.maa.org/press/periodicals/convergence/proofs-without-words-and-beyond

Fischbein E. and Engel, I. (1989). Psychological difficulties in understanding the principle of mathematical induction. In G. Vergnaud, J. Rogalski and M. Artigue (Eds.), *Proceedings of the 13th International Conference for the Psychology of Mathematics Education*, Volume I, 276–82, Paris: CNRS.

Giaquinto, M. (2007). *Visual Thinking in Mathematics: An Epistemological Study*. New York, NY: Oxford University Press.

Giardino, V. (2010). Intuition and visualization in mathematical problem solving. *Topoi*, 29(1):29–39.

Hanna, G. (2000). Proof, explanation and exploration: An overview. *Educational Studies in Mathematics*, 44(1):5–23.

Hanna, G. and Sidoli, N. (2007). Visualisation and proof: A brief survey of philosophical perspectives. *ZDM*, 39(1–2):73–78.

Harel, G. and Sowder, L. (1998). Students' proof schemes: Results from exploratory studies. In A. Schoenfeld, J. Kaput and E. Dubinsky (Eds.), *Research in Collegiate Mathematics Education III*, 234–83. Providence, RI: American Mathematical Society.

Jamnik, M. (2001). *Mathematical Reasoning with Diagrams*. Stanford, CA: CSLI Publications.

Kulpa, Z. (2009). Main problems with diagrammatic reasoning. Part 1: The generalization problem. *Foundations of Science*, 14(1–2):75–96.

Lange, M. (2009). Why proofs by mathematical induction are generally not explanatory. *Analysis*, 69(2):203–11.

Mancosu, P. (2005). Visualization in logic and mathematics. In P. Mancosu, K. F. Jørgensen and S. A. Pedersen (Eds.), *Visualization, Explanation and Reasoning Styles in Mathematics*, 13–30. Dordrecht, The Netherlands: Springer.

Movshovitz-Hadar, N. (1993). The false coin problem, mathematical induction and knowledge fragility. *Journal of Mathematical Behavior*, 12:253–68.

Nelsen, R. B. (1993). *Proofs without Words: Exercises in Visual Thinking* (No. 1). Washington, DC: Mathematical Association of America.

Nelsen, R. B. (2000). *Proofs without Words: Exercises in Visual Thinking* (No. 2). Washington, DC: Mathematical Association of America.

Nelsen, R. B. (2016). *Proofs without Words: Exercises in Visual Thinking* (No. 3). Washington, DC: Mathematical Association of America.

Pasch, M. (1882/1926). *Vorlesungen über neuere Geometrie*. Reprint of the 1926 edition with M. Dehn by Springer, 1976.

Peirce, B. (1881). Linear associative algebra. *American Journal of Mathematics*, 4(1): 1.

Rav, Y. (1999). Why do we prove theorems? *Philosophia Mathematica*, 7(1):5–41.

Relaford-Doyle, J. and Núñez, R. (2017). When does a 'visual proof by induction' serve a proof-like function in mathematics? In G. Gunzelmann, A. Howes, T. Tenbrink and E. J. Davelaar (Eds.), *Proceedings of the 39th Annual Conference of the Cognitive Science Society*, 1004–09, London: Cognitive Science Society.

Rota, G. C. (1997). The phenomenology of mathematical proof. In F. Palombi (Ed.), *Indiscrete Thoughts*, 134–50. Boston, MA: Birkhäuser.

Selden, A. and Selden, J. (2003). Validations of proofs considered as texts: Can undergraduates tell whether an argument proves a theorem? *Journal for Research in Mathematics Education*, 34(1):4–36.

Stylianides, G. J., Sandefur, J. and Watson, A. (2016). Conditions for proving by mathematical induction to be explanatory. *The Journal of Mathematical Behavior*, 43:20–34.

Weber, K. and Mejía-Ramos, J. P. (2014). Mathematics majors' beliefs about proof reading. *International Journal of Mathematical Education in Science and Technology*, 45(1):89–103.

Weber, K. and Mejía-Ramos, J. P. (forthcoming). An empirical study on the admissibility of graphical inferences in mathematical proofs. In A. Aberdein and M. Inglis (Eds.), *Advances in Experimental Philosophy of Logic and Mathematics*. London: Bloomsbury.

Zhen, B., Weber, K. and Mejía-Ramos, J. P. (2016). Mathematics majors' perceptions of the admissibility of graphical inferences in proofs. *International Journal of Research in Undergraduate Mathematics Education*, 2(1):1–29.

6

An Empirical Study on the Admissibility of Graphical Inferences in Mathematical Proofs

Keith Weber and Juan Pablo Mejía-Ramos

1 Introduction

Griffiths (2000) declared that "a mathematical proof is a formal and logical line of reasoning that begins with a set of axioms and moves through logical steps to a conclusion" (2). In our experience, Griffiths is providing a fairly conventional account of mathematical proof. As Doyle et al. (2014) commented, "Practicing mathematicians, it would seem from a casual survey, tend to understand their completed proofs in these terms." However, philosophers of mathematical practice have been critical of accounts such as Griffiths's because not many proofs that mathematicians actually produce satisfy this description. Published proofs usually do not explicitly invoke "a set of axioms" (e.g., Feferman, 1999) and are almost never written in a formal language (e.g., Rav, 1999). Nonetheless, proofs do seem to employ "a logical line of reasoning" where each step in the proof is a logical consequence of previous statements. But what constitutes "a logical line of reasoning"? What types of logical inference are acceptable in a mathematical proof?

The issue of what constitutes a valid logical inference is a difficult question. At a minimum, we believe a permissible step in a proof must provide the reader with rational grounds to believe that the new step is a logically necessary consequence of previous assertions. However, this begs the question of what constitutes these rational grounds. Formalist accounts typically describe valid rules of inferences as those that can be found by applying one of the explicit rules of inference in the formal system in which the proof is couched.[1] However, philosophers of mathematics find such a description unhelpful because many inferences in the proofs that mathematicians actually produce cannot be expressed in a formal language, at least not without seriously distorting the semantic content of the inference (e.g., Larvor, 2012).

In this chapter, we investigate mathematicians' perceptions of a particular kind of inference in a particular setting. Specifically, we shed light on what types of graphical inferences mathematicians find permissible in a real analysis proof. Following Larvor (in press), we examine mathematicians' reactions to *metrical graphical inferences* (i.e., inferences whose validity depends on the accuracy of the graph that was drawn and whose validity can be changed by minor deformations) and *nonmetrical graphical inferences* (i.e., inferences whose validity does not depend on a the accuracy of the graph and whose validity is not vulnerable to local deformations). The goal of this chapter is threefold. First, we demonstrate that most mathematicians reject metrical graphical inferences as impermissible in a real analysis proof. Second, we show that many mathematicians regard nonmetrical graphical inferences as permissible in a real analysis proof. Third, we illustrate how mathematicians collectively disagree on the permissibility of nonmetrical inferences.

2 Background literature

2.1 The status of visual reasoning in mathematical proof

Visual reasoning is widely accepted as a critical component of mathematical practice; empirical studies of mathematical practice demonstrate that mathematicians regularly construct and reason about diagrams while solving problems and writing proofs (Schoenfeld, 1985; Samkoff et al., 2012; Stylinou, 2002). In this sub-subsection, we do not address whether diagrammatic inferences *should* be allowed in a proof—a topic addressed by philosophers such as Brown (2010), Feferman (2012), and Kulpa (2009). Instead we address the permissibility issue of whether contemporary mathematicians regard diagrammatic inferences as appropriate in a proof in their practice.

While drawing inferences from diagrams is an indispensable tool for forming conjectures and generating proofs, many find diagrammatical inferences to be too unreliable to be included in a proof. Inglis and Mejía-Ramos (2009) synthesized the viewpoints of many philosophers and mathematicians in describing "the *common view* of the role that visual representations play in mathematics—i.e., that pictures may be useful heuristic tools which suggest ways of understanding proofs, but they are nevertheless inappropriate when it comes to providing unequivocal, reliable evidence to support a mathematical claim, let alone providing a proof" (101; emphasis in the original). Similarly, Brown (2010) wrote that "though not universal, the prevailing attitude is that pictures are really no more than heuristic devices; they are psychologically suggestive and pedagogically important—but they *prove* nothing" (25, emphasis in the original).[2] Feferman (2012) and Kulpa (2009) also believe that the impermissibility of diagrams is the *status quo* in contemporary mathematics, even though they think it ought not to be.

Not all philosophers agree that diagrammatic reasoning is impermissible in a proof. Aberdein (2009) described visual proofs as a proof*, where proof* was defined as "species of alleged 'proof,' where there is either no consensus that the method provides proof, or there is a broad consensus that it doesn't, but a vocal minority or an historical precedent which points the other way" (1). Hence, to Aberdein, visual proofs* that rely on diagrammatic inferences are accepted by at least *some* mathematicians. Tanswell (2016) remarked in passing that diagrammatic inferences are commonly accepted in proofs.[3] Larvor (in press) and Rav (1999) observed that diagrammatic inferences are both permissible and common in some mathematical domains, such as knot theory.

Our perspective is that the extent that mathematicians are willing to accept diagrammatic inferences is a descriptive question about mathematical practice. As it is a claim about the state of the world, it can and should be addressed by empirical study. In this chapter, we show that many mathematicians accept diagrammatic inferences in a proof in a real analysis context.

2.2 On two different types of graphical inferences

Our chapter builds on a recent paper by Larvor (in press). Larvor, a philosopher, takes for granted that mathematicians sometimes accept diagrammatic inferences as permissible in a proof, at least within the domains of Euclidean geometry and contemporary knot theory. Larvor's aim is to understand commonalities in the permissible diagrams in geometry and knot theory to better understand when diagrammatic inferences would be permissible in a proof. Larvor argued that in geometry and knot theory, diagrammatic inferences within a proof shared three features:

a. it is easy to draw a diagram that shares or otherwise indicates the structure of the mathematical object;
b. the information thus displayed is not metrical; and

c. it is possible to put the inferences into systematic mathematical relation with other mathematical inferential practices.

By "metrical," Larvor is indicating that the information read from the diagram is sensitive to local deformations of that diagram. Conversely, "nonmetrical" means that the information read from the diagram is not sensitive to local deformations of the diagram.

In this chapter, we explore graphical inferences that are a particular type of diagrammatic inference. By a graphical inference, we are referring to any instance in which an inference is drawn about a function based on the analysis of a Cartesian graph of that function. We distinguish between two types of graphical inferences and discuss how they relate to Larvor's analysis. Canonical definitions in undergraduate calculus and real analysis are typically expressed sententially, using a combination of logical syntax and natural language. For instance, we might say that "f is a **strictly increasing** real-valued function if for all real numbers a and b, if $a < b$, then $f(a) < f(b)$." While graphs may be valuable for motivating or illustrating canonical definitions, they are not part of the definition and the definition avoids direct reference to a graph. Further, while graphs might be useful for generating or understanding proofs about strictly increasing functions, the actual proofs in practice avoid direct appeal to such graphs.[4] Nonetheless, many calculus and real analysis concepts have *graphical interpretations*. For instance, a function f is a strictly increasing real-valued function if the graph of f goes up as you read it from left to right. Similarly, we can say a *strictly positive real-valued function* is a function whose graph remains above the x-axis, a real-valued function has a *root at 0* if its graph passes through the origin, a real-valued function is *even* if its graph is symmetric across the y-axis, and so on.

In the context of real-valued functions, we say that an individual is making a *graphical inference* if the individual concludes that a function has a particular property on the basis that the graph of that function satisfies the graphical interpretation of that property. As a simple hypothetical example, imagine a student was asked to prove that $x = \sin x$ has only one solution. The student might sketch a graph of $f(x) = x - \sin x$, and infer from this graph that 0 is a root of $f(x)$ (because the graph crosses the origin), that $f(x)$ is strictly increasing (because the graph goes up as you read it from left to right), and that $f(x)$ could not have a positive root (because the graph would have had to increase after $x = 0$ but then decrease to cross the x-axis again for a positive x). Each of the three student observations above were graphical inferences.

Using the example above, we can illustrate Larvor's (in press) distinction between metrical and nonmetrical inferences in the context of real analysis graphs. If a student inferred that 0 is a root of $f(x)$ (because the graph crosses the origin) and that $f(x)$ is strictly increasing (because the graph goes up as you read it from left to right), these would be instances of *graphical metrical inferences* as these judgments depend on the accuracy of the graph that was drawn. With a slight deformation of the graph, the $f(x)$ might have a root slightly different from 0, or have two roots in a small neighborhood around 0, or have been decreasing on some short interval. More generally, inferring that "A function f has property P" with the warrant that "the graph of f visually satisfies a graphical interpretation of P," is an example of a graphical metrical inference. Graphical metrical inferences do not generally satisfy Larvor's criteria (b). These inferences about $f(x)$ depend critically on the accuracy of the graph of $f(x)$. For instance, we could define $g(x) = (1 - 10^{100}) x - \sin x + (1/10^{100})$, and the graphs of $f(x)$ and $g(x)$ would be indistinguishable to the human eye (even if using graphic software), but $g(x)$ would not have a root at 0 and would not be strictly increasing.

Again, referring to the example above, if a student inferred that $f(x)$ could not have a positive root (because the graph would have had to increase after $x = 0$, but then decrease to cross the x-axis again for a positive x), this would be an example of a *graphical nonmetrical inference* as this inference would not be negated if the graph of $f(x)$ was drawn inaccurately. More generally, inferences of the form "since a function f has property P, the function f must necessarily have property Q" with the warrant that "one cannot construct a graph of a function satisfying the graphical interpretation of P while not also satisfying a graphical interpretation of Q as this would violate principles of two dimensional space," are instances of graphical nonmetrical inferences. We argue that graphical nonmetrical inferences would satisfy Larvor's criteria (b). If a student knew that $f(x)$ was strictly increasing and had a root at 0 (say the student noted that $f(x)$ was strictly increasing and verified $f(0) = 0$ by computation), then the student's inference would not depend on the accuracy of the graph of $f(x)$.

2.3 A previous study with undergraduate students

The data presented in this chapter builds upon a recent mathematics education study that we conducted with undergraduate mathematics students (Zhen et al., 2016). In that study, we observed university mathematics students who had recently graduated with a degree in mathematics or who had completed their junior year. Participants followed a "think aloud" protocol as they constructed proofs for seven calculus statements (e.g., "Prove that $\int_{-a}^{a} \sin^3 x\, dx = 0$ for any real number a"). While completing these tasks, students were given access to a computer with a graphing calculator application. We observed that students frequently made graphical inferences. Some of these inferences were graphical metrical inferences and some were graphical nonmetrical inferences.[5] The key result of our qualitative study was that participants generally did not think that graphical metrical inferences were appropriate for a proof. When participants drew a graphical metrical inference, they usually sought to justify this inference by other means. However, our participants usually acted as if graphical nonmetrical inferences were permissible in a proof. They would include the graphical nonmetrical inferences that they drew in the proofs that they wrote without seeking alternative justification or expressing any doubt about its acceptability. A subsequent survey with ninety university mathematics students who had completed a course in real analysis showed the same trends. The participants were shown proofs with a graphical metrical inference and asked if they thought the inference was acceptable; most participants answered that these graphical metrical inferences were not acceptable. When they were shown proofs with graphical nonmetrical inferences, most answered that the graphical nonmetrical inferences were acceptable. In this chapter, we report on the results when the same survey was sent to mathematicians.

3 Methods

3.1 Rationale

Within mathematics education research, scholars often seek to examine the mathematical practices of professional mathematicians and students of mathematics. We briefly elaborate on the reasons for studying these different populations. A common goal of mathematics education is to have students engage in *authentic mathematical practice* and to acquire the same beliefs, skills, and dispositions that professional mathematicians have (for instance, see Harel and Sowder, 2007; Schoenfeld, 1985; Sfard, 1998). Hence, one reason to study the practices of professional mathematicians is to better understand what authentic mathematical practice actually is (for further discussion, see Weber et al., 2014)[6] and thereby form appropriate goals of mathematics instruction. Two common reasons for studying students' mathematical practice are to identify aspects of students' practice that are in need of remediation and to better understand where students are at to form appropriate starting points for instruction. In tandem, studying both the practices of mathematicians and students allows the researcher to identify *discrepancies* between how mathematicians and students do mathematics. Reducing these discrepancies then becomes a goal of mathematics instruction.

In Zhen et al. (2016), we identified two aspects of university students' mathematical practice—that students generally did not find graphical metrical inferences to be permissible in a proof and that many students would accept graphical nonmetrical inferences in proof. In the current study reported in this chapter, we studied mathematicians' beliefs about graphical inferences in a pedagogical setting (the teaching of calculus and real analysis). Our aim was to see what the goals of instruction with regard to graphical inferences and proof should be and to identify discrepancies between students' and mathematicians' practices with graphical inferences and proofs. Entering this study, our hypothesis was that most mathematicians would adopt what Inglis and Mejía-Ramos (2009) referred to as the "common view"—that *all* graphical inferences would be inappropriate for a proof in an undergraduate setting. Because we sought to identify discrepancies between mathematicians and undergraduate students, we deliberately aimed to keep the methodology in the current study and our previous study with undergraduates as consistent as possible. Consequently, our description of the procedure largely mirrors the corresponding description in Zhen et al. (2016).[7]

3.2 The use of an internet study

Following the methodology employed by Inglis and Mejía-Ramos (2009), we collected data through the internet in order to maximize our sample size. Recent studies have examined the validity of internet-based experiments by comparing this type of studies with their laboratory equivalents (e.g., Kranz and Dalal, 2000; Gosling et al., 2004). The notable degree of congruence between the two methodologies suggests that, by following simple guidelines, internet data has comparable validity to more traditional data. We adopted the measures described in Inglis and Mejía-Ramos (2009) to ensure the validity of our data.

3.3 Participants

We recruited mathematicians to participate in this study as follows. Twenty-four secretaries from top-ranked mathematics departments in the United States[8] were contacted and asked to distribute an email to the mathematics faculty and doctoral students in their department. This email invited recipients to participate in our study and included a hyperlink that directed interested parties to the website of the study. Once in this website, participants were asked if they were doctoral students, held a degree in mathematics, or were in an "other" category. Through this process, we recruited eighty-four mathematicians who completed our experiment.

3.4 Procedure

Upon participating in the experiment, participants were randomly assigned to the real analysis or introductory calculus group. Participants in the real analysis group received the following instructions:

> We will ask you to read three mathematical statements and proofs. After carefully reading, please respond to the questions that follow as if the proofs are items on an exam in a **real analysis** class in your university. Two questions will be asked following each proof. The first question will ask you whether you think a step in the proof is sufficiently justified for an exam in a **real analysis** class. The second question will ask you whether you would take points off of the exam for the step.
>
> We will first provide you with an annotated and answered sample item to clarify the questions we are asking. Each proof will also be separated into steps so the reasoning is easier to follow.
>
> (The phrase "real analysis" was given in bold font and underlined on the actual webpage.)

The text for the calculus group was identical, except the "real analysis class" phrases were substituted with "first year calculus class."[9]

Next, participants were shown a worked example to illustrate the ideas of the experiment. A main aim with this worked example was to make participants aware that one could accept a step in the proof as permissible, even if the step is building on a previous step that was not acceptable. Participants were shown a sample proof of the claim that "$\frac{1}{x^4+x^2+2x+1}+1$ was positive for all real-valued x." Step 1 of the proof claimed that "my roommate told me that $\frac{1}{x^4+x^2+2x+1}$ was positive." In the worked example, step 1 was evaluated as not being an adequate justification in a calculus/real analysis class and one for which an instructor would deduct points. Even though the claim in Step 1 is true, we presumed that appealing to one's roommate was not a permissible justification. Step 2 stated that since $\frac{1}{x^4+x^2+2x+1}$ was positive and 1 was positive, $\frac{1}{x^4+x^2+2x+1}+1$ was positive. In the worked example, this step was considered acceptable, since if the participant treated step 1 as correct, step 2 is only using the accepted fact that the sum of two positive numbers is positive.

After participants indicated that they understood the task, they were presented with three proofs in a randomized order. Each proof contained a graphical metrical inference and a graphical nonmetrical inference (the proofs are presented in Appendix and discussed shortly). Participants were initially shown the theorem statement. They were then shown a purported proof of the theorem statement with a graphical metric inference shown in red and were asked the following two questions: "Do you think the argument in step x is an adequate justification for the claim that [claim made in step x] if the proof was written on an exam for a **real analysis** student." Next they were asked, "Would you take points off for the justification of the step highlighted in red if the proof was written on an exam for a **real analysis** class?" (The "real analysis" appeared in bold in the instructions. The calculus group had "first year calculus" printed in place of "real analysis." The variable "x" represented the step number where the graphical metrical inference was highlighted in red). This was then repeated with the graphical nonmetrical inference highlighted in red and the same two questions were asked.

3.5 Materials

There were three proofs used in this study. Each proof contained a graphical metrical and a graphical nonmetrical inference. The complete tasks are given in the Appendix. We illustrate our task with Proof 2, which purports to establish that the derivative of e^{-x^2} is odd. (We adapted this proof from Raman, 2003). The proof begins by presenting a graph of $f(x) = e^{-x^2}$. Step 1 in the proof is the graphical metrical inference, stating "We can see from the graph that $f(x)$ is symmetric across the y-axis." This is a metrical inference because if there were slight deformations of the graph of $f(x)$, the inference that $f(x)$ was even might no longer be true. Step 3 in the proof is a graphical nonmetrical inference that builds upon Step 1, claiming, "Thus for any point a, the tangent line of f at a and the tangent line of f at $-a$ will be mirror images of each other. Thus the slopes of these tangent lines will have the same magnitude but opposite signs." This is a nonmetrical inference because it is not vulnerable to deformations of the graphs on which it is based (i.e., so long as the graph retained the property of being symmetric across the y-axis, the inference would still hold).

In addition to the three proofs found in the Appendix, we added two other tasks to verify that our participants were taking the survey seriously, that there were at least some inferences that they would find permissible, and that there were at least some inferences that they would not countenance. The first task involved an inference that we believed was clearly valid and acceptable and the second task was based on an inference that was clearly invalid and unacceptable. In Proof 3, we highlighted an inference that we regarded as clearly justified in an adequate manner (an algebraic demonstration that the solutions to $12x^2 - 4x^3 = 0$ are $x = 0$ and $x = 3$). When participants evaluated Proof 3, in addition to the graphical metrical inference and the graphical nonmetrical inference, they were also asked if this transparently good inference was appropriate and whether they would take points off for this step in the proof. We presumed that if participants were taking our tasks seriously, the participant would judge this inference as permissible and would not take points off for this inference.

For a transparently bad inference, we created an alternative proof to Proof 1 that we evaluated as clearly inadequate (claiming $x = 0$ was the only solution to an equation by verifying that $x = 0$ was a solution). Prior to reading Proof 1, the participants read a proof consisting entirely of the transparently bad inference. The participants were then asked if this inference was appropriate and whether they would take points off for this step in the proof. We presumed that if participants were taking our tasks seriously, they would claim this inference was not appropriate and they would take points off for it.

4 Results

4.1 Evaluation of graphical metrical inferences and graphical nonmetrical inferences

In Table 6.1, we present the percentage of mathematicians who thought each of the eight inferences in the study was acceptable for a proof on an exam. In Table 6.2, we aggregate participants' judgments across the three graphical metrical inferences and the three graphical nonmetrical inferences.

We first observe that nearly all participants evaluated the transparently good and bad justifications in the manner that we intended. From Table 6.1, we see that 98 percent of the participants thought the transparently good justification was acceptable in a proof and only 2 percent of the participants judged the transparently bad inference to be acceptable, indicating that nearly all participants were willing to say at least some inferences were acceptable and at least some inferences were not.

Related-samples Wilcoxon signed-rank tests reveal that for both the real analysis group and the calculus group, participants found more nonmetrical

Table 6.1 Participants' judgments of acceptability of an inference by inference type

Condition	Good Inf. (%)	Bad Inf. (%)	Metrical inferences			Nonmetrical inferences		
			Proof 1 (%)	Proof 2 (%)	Proof 3 (%)	Proof 1 (%)	Proof 2 (%)	Proof 3 (%)
Real analysis ($N = 41$)	98	2	5	17	12	61	51	46
Calculus ($N = 43$)	98	2	14	28	16	74	56	58
Total ($N = 84$)	98	2	10	23	14	68	54	52

Table 6.2 Participants' aggregate judgments of the acceptability of graphical metrical inferences and graphical nonmetrical inferences

Condition	Metrical inferences (%)	Nonmetrical inferences (%)
Real analysis ($N = 41$)	11	53
Calculus ($N = 43$)	19	63
Total ($N = 84$)	15	58

inferences than metrical inferences within a proof to be acceptable ($p < .001$ for both comparisons). As one might predict, the calculus participants appeared somewhat more likely to accept both metrical inferences and nonmetrical inferences than the real analysis participants. However, we found no statistically significant difference between the real analysis and calculus participants regarding their judgments of either the metrical inferences (Mann–Whitney, $U = 748$, $p = .127$) or the nonmetrical inferences (Mann–Whitney, $U = 740.5$, $p = .190$).

Figure 6.1 presents a histogram displaying the distribution of participants according to the number of metrical and nonmetrical inferences they deemed acceptable in a proof. Figure 6.1 reveals that most participants (73 percent) did not accept any of the metrical inferences as permissible in a proof. Most participants (85 percent) accepted at least one of the three nonmetrical inferences as permissible and 31 percent of the participants accepted all three nonmetrical inferences as permissible.

Finally we note that there was no statistical difference in between the twenty mathematicians with PhD degrees and the sixty-three doctoral students (one participant did not enter in this demographic information) either in their evaluations of the metrical inferences (13 percent vs 16 percent, $p = .978$) or the nonmetrical inferences (66 percent vs 55 percent, $p = .235$). The main finding that we highlight in the discussion section is mathematicians' willingness to accept

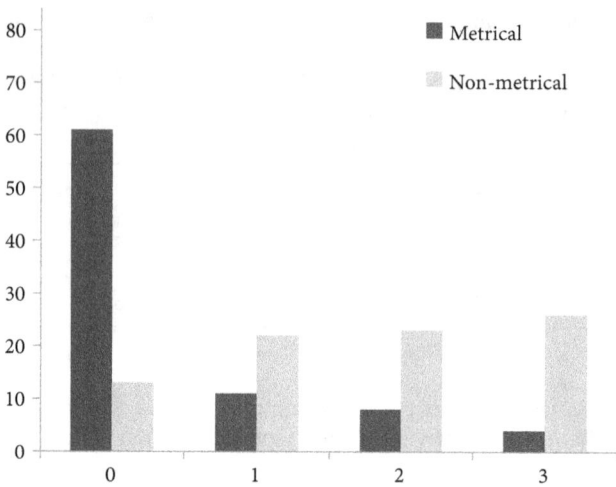

Figure 6.1 Distribution of participants according to the total number of graphical metrical and graphical nonmetrical inferences deemed acceptable across conditions.

nonmetrical inferences in proofs, a finding that was at least as pronounced for the mathematicians with PhD degrees in our sample as it was for the doctoral students.

4.2 On whether they would deduct points

Although the purpose of this chapter is not to investigate the grading practices of mathematicians,[10] we briefly summarize this data for the sake of completeness. In Table 6.3, we present the percentage of participants who thought a professor would take off points for each of the eight inferences in this study. In Table 6.4, we aggregate participants' judgments across the three graphical metrical inferences and the three graphical nonmetrical inferences. Tables 6.3 and 6.4 show trends similar to Tables 6.1 and 6.2. Most participants would deduct points for the transparently bad inference but few would deduct points for the transparently good inference. Related-samples Wilcoxon signed-rank tests demonstrated that both the real analysis participants and the calculus participants were more likely to take off points for the graphical metrical inferences than the graphical nonmetrical inferences ($p < .001$ in each case). Real analysis participants were more likely than the calculus participants to take points off for the graphical metrical inferences (Mann–Whitney, $U = 1152.5$, $p = .007$) but not for graphical nonmetrical inferences (Mann–Whitney, $U = 923$, $p = .695$).

Table 6.3 Participants' judgments of whether professor would take off points by inference

Condition	Good Inf. (%)	Bad Inf. (%)	Metrical inferences			Nonmetrical inferences		
			Proof 1 (%)	Proof 2 (%)	Proof 3 (%)	Proof 1 (%)	Proof 2 (%)	Proof 3 (%)
Real analysis ($N = 41$)	7	98	95	76	88	24	39	44
Calculus ($N = 43$)	7	95	84	56	72	21	37	37
Total ($N = 84$)	7	96	89	65	80	23	38	40

Table 6.4 Participants' aggregate judgments of metrical inferences and nonmetrical inferences

Condition	Metrical inferences (%)	Nonmetrical inferences (%)
Real analysis ($N = 41$)	86	36
Calculus ($N = 43$)	71	32
Total ($N = 84$)	78	34

5 Discussion and significance

5.1 Caveats and limitations

The biggest limitation of this study is that participants were evaluating proofs in an exam context in a calculus or real analysis setting, which differs in important respects from the proofs that they share with one another in mathematics journals. We can imagine some mathematicians being more likely to accept graphical inferences in a pedagogical setting as this could indicate conceptual understanding. We also think it is plausible that other mathematicians might be more reluctant to accept graphical inferences in a pedagogical setting if they thought the point of the undergraduate course was to help students learn how to represent their arguments using normative notation. Although grading student exams qualifies as a mathematical practice, this may not be the mathematical practice that philosophers have in mind when deciding what constitutes a proof. The extent that our results would generalize to other settings is an open research question. Consequently, we can only claim our results illustrate mathematicians' acceptance of nonmetrical inferences in pedagogical settings.

5.2 Relation to visual reasoning

In the beginning of this chapter, we made the case that the "common view" among scholars is that appeals to diagrams are not permissible in mathematical proofs. The data from this chapter challenge this view in the context of student-produced proofs at the undergraduate level. Collectively, the majority of nonmetrical inferences were accepted by the participants in this study. Further most participants (85 percent) accepted at least one nonmetrical inference as valid. This finding is consistent with the results from Weber and Czocher (in press), who found that 60 percent of the mathematicians in their study accepted a "Proof Without Words" (Nelsen, 1993) as a valid proof and 94 percent of the participants thought the proof would be valid in at least some contexts. In tandem, the results from these studies challenge the common view. Although many scholars claim that the mathematical community finds all diagrammatic inferences impermissible in a proof, the majority of mathematicians seem to accept at least some graphical inferences in proofs.

Based on his analysis of diagrammatic inferences across several mathematical domains, Larvor (in press) suggested three conditions that would be necessary for a diagrammatic inference to lead to epistemically secure knowledge, and hence appropriate in a proof. The inferences should be based on diagrams that are easy to draw, the inferences should be nonmetrical, and the inferences should be in systematic alignment with other mathematical practices. Our empirical data support Larvor's philosophical analysis in two respects. First, Larvor would find graphical metrical inferences as unable to secure mathematical knowledge and hence impermissible in a proof. The mathematicians in our sample agreed, rejecting most such inferences. Second, the issue of whether graphical inferences were metrical or nonmetrical appeared to be a critical factor in mathematicians' evaluations. Consistent with Larvor's analysis, mathematicians were much more likely to accept nonmetrical inferences than metrical ones.

5.3 Distinguishing between regularities and normative practices

When we conducted this study, our hypothesis was that mathematicians would find *all* types of graphical inferences to be inappropriate in a real analysis setting. We formulated our hypothesis due to our inspection of common real analysis texts and solution manuals. In Zazkis et al. (2016), we reported that real analysis textbooks usually do not accompany proofs with figures; when diagrams are present, no steps in the proof are justified explicitly or implicitly with appeal to the diagram. Likewise, student solution manuals simply include no reference to diagrams. Based on these findings and other researchers' analyses of mathematics textbooks (e.g., Raman, 2004), we concluded that graphical inferences were not permissible in a real analysis proof. The data in this chapter strongly suggest that our conclusion was wrong.

We believe our error should serve as a warning to avoid a particular type of unwarranted inference. Specifically, just because most proofs satisfy some property, this does not imply that a justification must satisfy this property to count as a proof. Similarly, just because most proofs lack a particular property does not mean that a justification that satisfies this property cannot count as a proof. As a trivial illustration, we have never seen a textbook proof or a published proof that contained a valid string of computations that were irrelevant to the proof. However, we would not want to say that any justification with a pointless valid string of calculations could never be a proof (even though the inclusion of such a string may negatively affect mathematicians' appraisal of the proof and perhaps their willingness to publish it).

5.4 Comparing mathematicians to university mathematics students and implications for education

We conducted this study with the aim of establishing a discrepancy between mathematicians' and university mathematics students' expectations for what inferences should be acceptable within a proof. It turned out that the students' collective beliefs about proof better aligned with mathematicians than our initial beliefs! When we administered an analogous survey to university mathematics students (Zhen et al., 2016), students' responses were similar to mathematicians' responses reported in this chapter. The students rejected most graphical metrical inferences and accepted the majority of graphical nonmetrical inferences. Collectively, the students found both metrical and nonmetrical inferences to be more permissible in a calculus context than a real analysis context, but this effect was not statistically significant.

This has an important consequence for mathematics education. In undergraduate mathematics education, researchers have observed that university mathematics students perform poorly when they are asked to prove statements (e.g., Iannone and Inglis, 2010; Weber, 2001). Mathematics educators often try to remediate this situation by aligning students' standards of conviction with those held by mathematicians (e.g., Brown, 2014; Harel and Sowder, 2007) so that students can draw valid inferences in the proofs that they produce. The results of our empirical studies with mathematicians and university mathematics students suggest that students collectively have the right standards. This is a result that we have found with other university students who have had training in proof-oriented mathematics (Weber, 2010; Weber, Lew, and Mejía-Ramos, in press; see also Iannone and Inglis, 2010). While *a minority* of university mathematics students gain high-levels of conviction from example-based reasoning and graphical metrical inferences, most do not. We would therefore urge researchers to move away from finding ways to amend students' deficient standards for gaining conviction, and toward developing the cognitive skills to make *productive* inferences when writing a proof and recognizing errors in logic when reading them.

5.5 On variation in mathematicians' evaluations

Suppose we were to ask the question: does the mathematical community find graphical nonmetrical inferences permissible in a real analysis proof in a pedagogical context? Our data suggest that this question does not have an answer, at least not a binary one. For each of the graphical nonmetrical inferences, a large number of our participants (46 percent–61 percent) evaluated the inference as permissible and a large number of our participants (39 percent–54 percent) evaluated the inference as impermissible. This is consistent with other empirical studies that we have conducted (Inglis et al., 2013; Weber, 2008; see also Inglis and Alcock, 2012; Weber and Czocher, in press), which have found variation in mathematicians' acceptance of given proofs, and with a broader literature showing wide variation in mathematical practice (for a review, see Weber et al., 2014). Larvor (in press) noted that while he has proposed necessary conditions for a graphical inference to rigorously secure knowledge, more research is needed to find sufficient conditions. The results in this chapter offer a challenge to this research program: what is sufficient for one mathematician is likely not sufficient for another.

Notes

1 For instance, Morash (1987) described a proof as "a series of statements, each of whose validity is based on an axiom or a previously proved theorem" (p.149).
2 Note that Inglis and Mejía-Ramos (2009) did not endorse the "common view" and Brown (2010) disagreed with such view.
3 Tanswell (2016) commented that for the proofs that mathematicians actually produce "steps in these proofs can rather be leaps and invoke the background

knowledge of your target audience, the semantic understanding of the terms being employed, visualisation, diagrams and topic-specific styles of reasoning" (p. 161).

4 Or so we presume. We found that references to graphs in real analysis textbooks and solution manuals to these textbooks were virtually absent (Zazkis, Weber, & Mejía-Ramos, 2016). However, we have not systematically analyzed articles in real analysis journals, colloquiums on these topics, and so on.

5 In Zhen et al. (2016), we referred to graphical metrical inferences as "graphical perceptual inferences" and graphical nonmetrical inferences as "graphical deductive inferences." We adopt Larvor's (in press) improved terminology to fit with the themes of this edited volume.

6 In Weber et al. (2014), we describe how some commonly held assumptions about mathematical practice in mathematics education, such as mathematicians not gaining conviction in mathematical theorems and proofs based on the word of an authoritative source, are actually not descriptive of mathematicians' practice.

7 To avoid misinterpretation, the undergraduate student data was presented in Zhen et al. (2016). The data with mathematicians is presented for the first time in this chapter.

8 As ranked by the USNews.com "Best Graduate Schools" list of "top mathematics programs."

9 The distinction between real analysis and calculus was included in this study because we made the same distinction in Zhen et al. (2016). This factor is of no theoretical relevance for the current chapter.

10 Our original aim was to highlight a discrepancy between these mathematicians' views and the views of undergraduate mathematics students from Zhen et al. (2016) on what types of inferences would be penalized in a calculus or real analysis proof. However, as we discuss in the section "Discussion and significance," no such discrepancy existed.

References

Aberdein, A. (2009). Mathematics and argumentation. *Foundations of Science*, 14:1–8.

Brown, J. R. (2010). *Philosophy of Mathematics: An Introduction to the World of Proofs and Pictures*. New York, NY: Routledge.

Brown, S. A. (2014). On skepticism and its role in the development of proof in the classroom. *Educational Studies in Mathematics*, 86(3):311–35.

Doyle, T., Kutler, L., Miller, R. and Schueller, A. (2014). Proofs without words and beyond. *Convergence*. Retrieved December 6, 2018 from http://www.maa.org/press/periodicals/convergence/proofs-without-words-andbeyond.

Feferman, S. (1999). Does mathematics need new axioms? *The American Mathematical Monthly*, 106(2):99–111.

Feferman, S. (2012). And so on…: reasoning with infinite diagrams. *Synthese*, 186(1):371–86.
Gosling, S. D., Vazire, S., Srivastava, S. and John, O. P. (2004). Should we trust web-based studies? A comparative analysis of six studies about internet questionnaires. *American Psychologist*, 59:93–104.
Griffiths, P. A. (2000). Mathematics at the turn of the millennium. *American Mathematical Monthly*, 107:1–14.
Harel, G. and Sowder, L. (2007). Toward comprehensive perspectives on the learning and teaching of proof. In F. K. Lester (Ed.), *Second Handbook of Research on Mathematics Teaching and Learning*, 805–42. Greenwich: Information Age Publishing.
Iannone, P. and Inglis, M. (2010). Self-efficacy and mathematical proof: Are undergraduates good at assessing their own proof production ability? In *Proceedings of the 13th Conference for Research in Undergraduate Mathematics Education*, February 25–28, 2010, Raleigh, North Carolina.
Inglis, M. and Alcock, L. (2012). Expert and novice approaches to reading mathematical proofs. *Journal for Research in Mathematics Education*, 43(4):358–90.
Inglis, M. and Mejía-Ramos, P. (2009). On the persuasiveness of visual arguments in mathematics. *Foundations of Science*, 14:97–110.
Inglis, M., Mejía-Ramos, J. P., Weber, K. and Alcock, L. (2013). On mathematicians' different standards when evaluating elementary proofs. *Topics in Cognitive Science*, 5:270–82.
Kranz, J. and Dalal, R. (2000). Validity of web-based psychological research. In M. Birnbaum (Ed.), *Psychological Experiments on the Internet*, 35–60. San Diego: Academic.
Kulpa, Z. (2009). Main problems with diagrammatic reasoning. Part 1: The generalization problem. *Foundations of Science*, 14:75–96.
Larvor, B. (2012). How to think about informal proofs. *Synthese*, 187(2):715–30.
Larvor, B. (in press). From Euclidean geometry to knots and nets. To appear in *Synthese*.
Morash, R. P. (1987). *Bridge to Abstract Mathematics: Mathematical Proof and Structures*. New York, NY: Random House.
Nelsen, R. B. (1993). *Proofs without Words: Exercises in Visual Thinking*. Washington, DC: MAA.
Raman, M. (2003). Key ideas: what are they and how can they help us understand how people view proof? *Educational Studies in Mathematics*, 52:319–25.
Raman, M. (2004). Epistemological messages conveyed by three high-school and college mathematics textbooks. *The Journal of Mathematical Behavior*, 23(4):389–404.
Rav, Y. (1999). Why do we prove theorems? *Philosophia Mathematica*, 7(3):5–41.
Samkoff, A., Lai, Y. and Weber, K. (2012). On the different ways that mathematicians use diagrams in proof construction. *Research in Mathematics Education*, 14(1):49–67.
Schoenfeld, A. H. (1985). *Mathematical Problem Solving*. Orlando: Academic.
Sfard, A. (1998). On two metaphors for learning and the dangers of choosing just one. *Educational Researcher*, 27(2): 4–13.

Stylianou, D. A. (2002). On the interaction of visualization and analysis – The negotiation of a visual representation in problem solving. *The Journal of Mathematical Behavior*, 21(3):303–17.

Tanswell, F. S. (2016). Saving proof from paradox: Gödel's paradox and the inconsistency of informal mathematics. In H. Andreas and P. Verdée (Eds.), *Logical Studies of Paraconsistent Reasoning in Science and Mathematics*, 159–74. Cham, Switzerland: Springer.

Weber, K. (2001). Student difficulties in constructing proofs: The need for strategic knowledge. *Educational Studies in Mathematics*, 48(1):101–19.

Weber, K. (2008). How mathematicians determine if an argument is a valid proof. *Journal for Research in Mathematics Education*, 39:431–59.

Weber, K. (2010). Mathematics majors' perceptions of conviction, validity, and proof. *Mathematical Thinking and Learning*, 12(4):306–36.

Weber, K. and Czocher, J. (in press). On mathematicians' disagreement on what constitutes a proof. To appear in *Research in Mathematics Education*.

Weber, K., Inglis, M. and Mejía-Ramos, J. P. (2014). How mathematicians obtain conviction: Implications for mathematics instruction and research on epistemic cognition. *Educational Psychologist*, 49:36–58.

Weber, K., Lew, K. and Mejía-Ramos, J. P. (in press). Using expectancy value theory to account for students' mathematical justifications. To appear in *Cognition and Instruction*.

Zazkis, D., Weber, K. and Mejía-Ramos, J. P. (2016). Bridging the gap between graphical arguments and verbal-symbolic proofs in a real analysis context. *Educational Studies in Mathematics*, 93(2):155–73.

Zhen, B., Weber, K. and Mejía-Ramos, J. P. (2016). Mathematics majors' perceptions of the admissibility of graphical inferences in proofs. *International Journal of Research in Undergraduate Mathematics Education*, 2(1):1–29.

Appendix—Tasks used in study

Proposition 1. *Prove the only real solution to the equation $x^3 + 5x = 3x^2 + sin(x)$ is $x = 0$.*

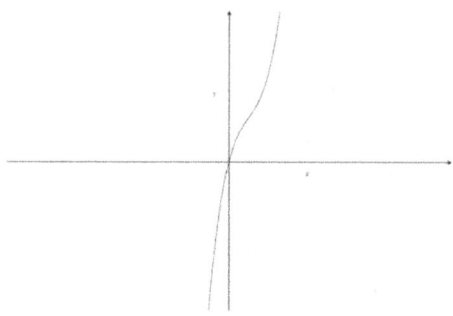

$$f(x) = x^3 - 3x^2 + 5x - sin(x)$$

Proof.
Step 1. We can rewrite the given equation as $x^3 - 3x^2 + 5x - sin(x) = 0$. Then it suffices to show that $f(x) = x^3 - 3x^2 + 5x - sin(x) = 0$ if and only if $x = 0$.

Step 2. First, we know that $x = 0$ is a solution since $f(0) = 0^3 - 3*0^2 + 5*0 - sin(0) = 0$. So we show that $x = 0$ is the only solution.

Step 3. Given the graph of $f(x)$ above, we see that $f(x)$ is strictly increasing.

Step 4. Because $f(x)$ is strictly increasing, $f(x)$ does not have a positive root, since it will have to come back down or remain flat which contradicts the fact that $f(x)$ is strictly increasing.

Step 5. Similarly, f(x) does not have a negative root. Thus $x = 0$ is the only solution. □

STEP 3 is the graphical metrical inference
STEP 4 is the graphical nonmetrical inference

Proposition 1. *Prove that the derivative of $f(x) = e^{-x^2}$ is odd.*

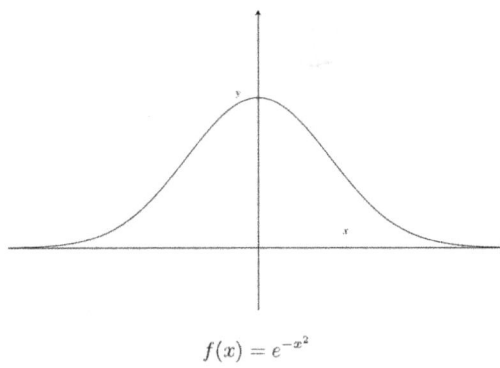

$$f(x) = e^{-x^2}$$

Proof.

Step 1. From the graph, we see that $f(x)$ is symmetric across the y-axis.

Step 2. For any a, the slope of the tangent line of f at a is $f'(a)$.

Step 3. By step 1, $f(x)$ is symmetric across the y-axis. Thus, for any point a, the tangent line of f at a and the tangent line of f at $-a$ will be mirror images of each other. Thus the slopes of these tangent lines will have the same magnitude but opposite signs.

Step 4. Thus, from step 2 and step 3, we get $f'(a) = -f'(-a)$ for any a and hence f' is odd, as desired. □

STEP 1 is the graphical metrical inference
STEP 3 is the graphical nonmetrical inference

Proposition 1. *Prove that the equation $4x^3 - x^4 = 30$ has no real solutions.*

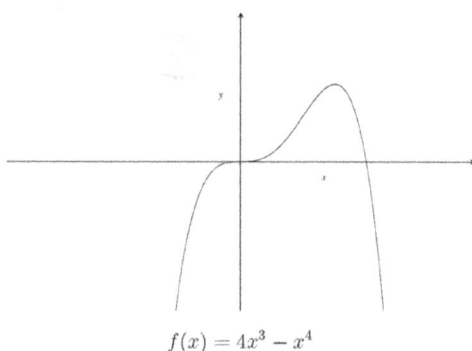

$$f(x) = 4x^3 - x^4$$

Proof.
Step 1. Let $f(x) = 4x^3 - x^4$. Since $f(x)$ is polynomial function, $f(x)$ is continuous.

Step 2. From the graph above, we see that $f(x)$ tends to $-\infty$ as x tends to $-\infty$.

Step 3. Similarly, $f(x)$ tends to $-\infty$ as x tends to ∞.

Step 4. By steps 2 and 3, $f(x)$ tends to $-\infty$ as x tends to $-\infty$ and x tends to ∞. Therefore, $f(x)$ must be bounded. If not, $f(x)$ must have a vertical asymptote, which implies $f(x)$ is not continuous.

Step 5. Taking the derivative of $f(x)$ we get $f'(x) = 12x^2 - 4x^3$.

Step 6. Setting $f'(x) = 0$ gives $12x^2 - 4x^3 = 0$. This can be simplified to $4x^2(3-x) = 0$. Hence, a solution to this equation occurs when $4x^2 = 0$ or $3 - x = 0$ and so $x = 0$ or $x = 3$. Thus $f'(x) = 0$ when $x = 0$ and $x = 3$. Thus $f(x)$ has critical points $x = 0$ and $x = 3$.

Step 7. Since $f(x)$ is bounded above, one of these critical points must be a global maximum. Since $f(0) = 0$ and $f(3) = 27$, $f(0) < f(3)$, and thus $f(3)$ is a global maximum. So $f(x) < f(3) < 30$ for all $x \in \mathbb{R}$, the equation $4x^3 - x^4 = 30$ has no real solutions. $\square$

STEP 2 is the graphical metrical inference
STEP 4 is the graphical nonmetrical inference
STEP 6 is the transparently good inference

7

Does Anyone Really Think That ⌜φ⌝ Is True If and Only If φ?

Robert Barnard and Joseph Ulatowski

The biconditional schema form: ⌜φ⌝ is true if and only if φ is a common feature of various philosophical attempts to characterize truth.[1] Sometimes this principle is arrived at by reflection upon "'φ' is true" as asserting nothing other than "φ" (cf. Frege, 1919/1956, Ramsey, 1927). Other theorists arrive at the principle as a generalization by reasoning from particular instances of the biconditional (e.g., Horwich, 1990/1998) to the biconditional schema. However, biconditionals of this form are most closely associated with Alfred Tarski. For the remainder of the chapter, let:

$$ES_{def} = ⌜φ⌝ \text{ is true if and only if } φ.[2]$$

Orthodox interpretations of Tarski's semantic conception of truth (1936/1983, 1944) combine a formal and generalizable schema with the material adequacy condition for truth. The regimentation of the concept of truth is accomplished by specifying the structure of the languages for which truth is defined and by specifying a formal criterion of material adequacy in terms of whether all instances of the equivalence,

Convention-T: X is true-in-$\mathscr{L}$ if, and only if p

follow from the proposed definition. "X" is the name or description of a sentence in $\mathscr{L}$, an object language, and p is a translation of the object language sentence into meta-language. For Tarski, material adequacy allows for the analysis of sentences into sub-sentential constituents, and it dissolves correspondence relations into appropriate semantic components: names refer to or denote objects and predicates apply to or are satisfied by objects.[3] According to Tarski, any theory of truth must entail, for any sentence X in a given language, a sentence of the form of Convention-T. Tarski makes it clear that Convention-T is supposed to conform to the intuition behind the concept of truth, which he refers to as the "classical," "old notion," or "Aristotelian conception" of truth (cf. Tarski, 1936/1983: 155, 1944: 342f.). From Convention-T, there follows instances or what we call *practical variants* of ES that each constitute a partial definition of the concept of truth, for example, ⌜Snow is white⌝ is true if and only if snow is white.

While Tarski's formal semantic conception of truth, and arguments leading to it, may be too technical for the nonphilosopher to fully comprehend, they should not have trouble embracing expressions that characterize instances of ES. In fact, Tarski held that the partial or practical characterization of truth exemplified by instances of ES was supposed to capture the content of the "everyday" or "commonsense" notion of truth, a view of truth employed and upheld by language users.

One underappreciated facet in how we ought to understand Tarski's position turns on whether empirical work examining the nonphilosopher's notion of truth ought to inform our understanding of Tarski's semantic conception of truth (cf. Barnard and Ulatowski, 2016; Chapman, 2018; Ulatowski, 2016, 2017). A historical precursor here is the empirical studies of Arne Næss (1938a, 1938b, 1953). Næss's contribution to the Norwegian school of "empirical semantics" prefigures empirical studies completed under the auspices of the contemporary movement known as "Experimental Philosophy" (Alexander, 2012; Knobe and Nichols, 2008, 2014; Sytsma and Livengood, 2015). This chapter employs empirical means to assess whether Tarski's attempt to formally characterize truth has succeeded in capturing the nonphilosopher's, or commonsense, notion of truth that he had claimed was the target of his analysis (cf. Tarski, 1936/1983: §1, 1944: Part II, 1969).

In this chapter, we provide evidence that while a majority of nonphilosophers judge that a given statement prefixed by "It is true that . . ." expresses the same thing as the statement itself in some cases, they tend to judge that when a free variable, for example, "p," stands in for any sentence, that expression states something different than the expression created by prefixing p with "It is true that . . ." or by predicating p with "is true."[4] To get to that conclusion, we propose to survey examples of relevant historical (and recent) empirical results and then discuss why these results should be addressed through empirical studies. We discuss Næss's empirical findings as reported not only by Næss himself but by

Tarski. Then we turn our attention to several empirical studies we performed exploring instances of ES. We argue that these findings collectively suggest: (a) that acceptance of ES can vary according to the conceptual terrain (especially in light of Næss's findings) and (b) that acceptance of such expressions differs according to whether the case is considered in an abstract or a practical (i.e., particular) format.

1 The empirical in the semantic conception of truth

The project of attempting to empirically assess whether ordinary people accept either Convention-T or instances of ES has a long history. The first mention of such an empirical project comes from Næss (1938a): "[Professional Formulation] 148, "'p' is wahr, wenn p' of Tarski received much criticism as well as appraisal." A footnote to this sentence adds,

> Special questionnaires were used to establish the willingness to substitute "p" for "p is true." Suitable examples were listed in the [questions]. The results cannot be stated in a few words and must be omitted in this work.—The [formulation 148] (Tarski) is not to be identified with the so-called "semantic notion of truth." To construct this notion the method of formulation is essential. There is, however a tendency to look at [formulation 148] as a definition of nonformal truth. It was therefore included in our lists. (Næss, 1938a: 148)

Unfortunately, it appears that Næss never published these results in full, but, as Tarski pointed out, they were discussed at the Paris Congress of 1935 and Næss at least gestured at them in his work (especially, Næss, 1938a). We know very little about Næss's early studies of Tarski's formulation that were completed prior to his 1938 monograph, other than what we learn from Tarski. Tarski claims that he was "by no means surprised to learn" that Næss's results show only 15 percent of people accept the "philosophical formulation" of the classical correspondence conception of truth and 90 percent of people agreed with the "plain word" expression ("It is snowing" is true, if and only if, it is snowing) of the "same conception" of truth (Tarski, 1944: 360).

While his work on a formal semantic conception of truth has had a lasting impact upon theories of truth, Tarski was open to the possibility that the nonphilosopher's notion of truth may be studied via empirical means, that is, a "scientific survey questionnaire."

> I happen to believe that the semantic conception does conform to a very considerable extent with the common-sense usage although I readily admit I may be mistaken. What is more to the point, however, I believe that the issue raised can be settled scientifically, though of course not by a deductive procedure, but with the help of the statistical questionnaire method. As a matter of fact, such research has been carried on, and some of the results have been reported at congresses and in part published. (Tarski, 1944: 360)

According to Tarski, we need not restrict ourselves to our own intuitive judgments about truth if we want to appreciate what the contours of the ordinary person's view of truth are. By looking at and considering the results of empirical studies an important fixture in theorizing about truth, we need not be put into a position of rejecting or calling into question the formal semantic conception of truth. For nothing about the nonphilosopher's view of truth could be introduced that causes us to reconsider the viability of the semantic conception. Even if some empirical data were found to generate some doubt, philosophers could easily explain away anomalous results in the empirical data.

There is a reasonable amount of evidence supporting the view that Tarski conceived of the nonphilosopher's notion of truth as different from Convention-T. Indeed, in §1 of Tarski (1936/1983), Tarski offers an account of why the formal characterization of "true sentence" must be distinct from the nonphilosopher's colloquial view. He writes,

> *The very possibility of a consistent use of the expression "true sentence" which is in harmony with the laws of logic and the spirit of everyday language seems to be very questionable, and consequently the same doubt attaches to the possibility of constructing a correct definition of this expression.* (Tarski, 1936/1983: 165, original italics)

Tarski distinguishes between the formal correctness condition of truth and the "spirit" of how nonphilosophers use the term "true," but does not summarily dismiss the everyday or commonsense usage of the term.

While there is no doubt that Convention-T provides a framework for truth in a formal language, Tarski's further aim was to ensure that the semantic theory was "in harmony with" the use of truth in natural language. Tarski recognized, however, that there are difficulties faced by attempts to clarify the truth-concept using only the resources of ordinary language because "like other words from our everyday language, [true] is certainly not unambiguous" (Tarski, 1944: 342). The extension of Tarski's concept of truth is given by the axioms derived from ES, for example, *"snow is white" is true, if and only if snow is white*. ES is used to assess the material adequacy of the inductive definition of truth which lies at the heart of any realization of Tarski's semantic conception. For any instance of ES, Tarski must have realized that it would not suffice for a general definition of truth because, for any given instance, it could only speak to the content of that particular example. We could imagine innumerable sentences of the following sort:

> "Snow is white" is true if and only if snow is white.
> "Chata is Choctaw" is true if and only if Chata is Choctaw.
> "Abel Tasman sighted New Zealand in 1642" is true if and only if Abel Tasman sighted New Zealand in 1642.
> ⋮

We cannot infer Tarski's semantic theory of truth, including the formal correctness and material adequacy conditions, from instances of ES alone. For this reason, Tarski began with a formally defined semantic conception of truth.

Convention-T is understood to offer a material adequacy condition for Tarski's semantic definition of truth—accordingly, a correct theory of truth on Tarski's view is supposed to entail all instances of ES. Instances of ES, while not a formal part of Tarski's semantic conception, are easily comprehended by users of natural language. (It seems to go without saying that language users are competent enough to understand the instance "Harvard University was established in 1636" is true if and only if Harvard University was established in 1636.) Thus, we would expect that users of natural language would have a relatively straightforward positive intuitive response to particular instances of ES; they would find them intuitively acceptable.

2 Næss's 1953 experimental work

In a 1953 study, Næss reported his work on intuitive responses to questions asking about Tarski's semantic conception of truth (Næss, 1953: 5–7).[5] The data reported by Næss in the 1953 studies are consistent with the results he reported in his earlier work. The ordinary conception of truth, though, deeply fragmented along multiple lines of inquiry tended to show that people agree with instances of ES. The aim of this section is to provide an overview of the results of that 1953 study.

The study involved 130 Norwegian respondents who were enrolled in sophomore level courses at the University of Oslo. The questionnaire consisted of two parts: (i) a general overview of synonymy with a few examples and a set of criteria participants were recommended to employ and (ii) a list of sentences where respondents were asked whether any of the sentences were expressive of the same assertion.

Part (i) contained a unique training mechanism respondents were encouraged to use to answer questions of synonymy. Næss provided the following criteria for respondents:

> A sentence A is for you expressive of a different assertion from that of another sentence B, if and only if you can imagine possible (but perhaps not actual) circumstances (conditions, existing state of affairs) of such a kind that if they were present you would accept A as warranted but reject B, or vice versa.
>
> A sentence A expresses for you one and the same assertion as another sentence B, if and only if you cannot imagine such circumstances, that is, if you under all conditions whatsoever either would accept both A and B as warranted, or reject both A and B. (Næss, 1953: 38)

The wording of the criteria may sound quirky to the twenty-first-century ear, but there seems little reason for us to be concerned. First, Næss's criteria were aimed at an audience of Norwegian undergraduates of the early-to-mid-twentieth century. Their vernacular was quite distinct from our own, and Næss accommodated the colloquial language popular among young Norwegian undergraduates. Second, the criteria were translated from Norwegian to English in Næss's article. Given that there are significant linguistic differences between the two languages, Norwegian and English, that may contribute to its awkwardness.[6]

The presentation of these guidelines occurred immediately before part (ii) of the questionnaire, which involved respondents choosing whether a pair of sentences was judged to be synonymous. A series of sentences were labeled A through G and respondents were asked, "Are some of the following sentences for you expressive of the same assertion, and if that is the case, which ones": (Næss, 1953: 39). The labeled sentences had the following general structure:

A. p (or not p).
B. It is true (or not true) that p.
C. It is perfectly certain (or not perfectly certain) that p.
D. It is extremely probable (or not extremely probable) that p.
****

E. It is true that not-*p*.
F. It is extremely probable that not-*p*.
G. It is perfectly certain that not-*p*.[7]

Næss hypothesized that "there is a more pronounced tendency to affirm the synonymity of 'p' and 'It is true that p' in case 'p' is a sentence expressing a (supposed) matter of fact than in case 'p' expresses . . . a theory" (Næss, 1953: 16).

2.1 A first comparison case

Let's focus on the content of Questions 1A and 1B and Question 7A and 7B. In each of the questions, *p* was replaced with:

Question 1: There is at least one copy of the Bible in the University Library.
Question 7: [A formulation of Darwin's selection theory, e.g.] Darwin's selection theory explains the evolution of populations through change in heritable traits over time.[8]

Generally, the declarative sentence of question 1 is an empirical matter of fact that states there is a copy of the Bible in the university library. The sentence of question 7, on the other hand, involves a scientific theory, namely Darwin's evolutionary theory.

As Næss had hypothesized, affirmation of synonymy seemed to vary according to whether the sentence was an empirical matter of fact or a statement about a scientific theory. A majority of people responded that the two expressions 1A and 1B were the same, while a majority of respondents who received a statement of (7A) "the selection theory of Darwin" and (7B) "Darwin's selection theory is true" tended not to affirm that the two statements were synonymous (see Table 7.1.). Næss never compared the data of these two questions using statistical analysis. Since the raw data are published in Næss's book, we did and found that the results are statistically significant (χ^2 (1) = 40.9116, $p < .01$).

Table 7.1 Comparison of responses to Q1 and Q7

	Q1 (%)	Q7 (%)
Synonymy affirmation	87	44
Synonymy denial	13	56

2.2 Discussion of the first comparison case

Næss's study shows that people seem to think about a partial and informal rendering of Tarski's semantic conception of truth differently, depending upon features of the expressions contained in ES. While it would be much too hasty for us to conclude that nonphilosophers reject the semantic conception of truth, Næss's data give us reason to believe that some characteristic of the sentences tend to inform people's intuitive responses. Næss writes,

> In 7A–7D, the term "theory" reminds the readers of the kind of general and abstract sentence they have to consider. There is a tendency in relation to such sentences to find the term "true" to be inapplicable or somehow too "strong."
> (Næss, 1953: 13)

Næss suggests that nonphilosophers tend to hedge away from affirming the truth of scientific theories because for them a theory amounts to nothing more than an unconfirmed hypothesis. But a scientific theory is not an unconfirmed hypothesis easily refuted by observing data that contradict the opinion; rather, a scientific theory, particularly one as well documented as Darwin's, proposes a testable hypothesis where follow-on tests are able to confirm or disconfirm the hypothesis. People, so it seems from the above data, lack a proper understanding of what a scientific theory is. If they had a proper understanding of what a scientific theory is, at least with respect to the nature of repeatable experimental results under strictly controlled conditions, then they would affirm that 7A and 7B are the same. People's failure to affirm that a sentence describing Darwin's evolutionary theory is synonymous with that same sentence when it's prefixed by "It is true that . . ." might have more to do with the specific content of 7 than it has to do with scientific theorizing generally. The question is whether people do this for all scientific theories or just a select few.

2.3 Second comparison case

If people were sensitive to the specific content of 7, that is, Darwin's theory of evolution, because of background religious belief or a general distrust of the scientific enterprise, then people would likely affirm that formulations A and B are synonymous even in cases of the content having to do with an alternative scientific theory. In question 6, using the same methods as summarized above,

Table 7.2 Comparison of responses to Q6 and Q7

	Q6 (%)	Q7 (%)
Synonymy affirmation	85	44
Synonymy denial	15	56

Næss asked study participants whether the following two sentences were synonymous:

Question 6A: There is an ether that oscillates in accordance with the following laws—(here you may imagine a complete formulation of "the wave theory of light" or closely related theories).

Question 6B: It is true that there is an ether that oscillates in accordance with the following laws—(here you may imagine a complete formulation of "the wave theory of light" or closely related theories).

When we contrast the data set of Question 6 with Question 7, we find an asymmetry similar to the one when we review the data for Questions 1 and 7 (see Table 7.2.).

Just as with the previous case, we compared the empirical results, which yielded a statistically significant difference (χ^2 (1) = 36.707, $p < .001$).

2.4 Discussion of the second comparison case

There seems to be something peculiar having to do with Darwin's evolutionary theory that doesn't seem to apply to other scientific theories, such as the wave theory of light, in testing people's intuitive responses to practical variants of Tarski's semantic conception of truth. It appears that people tend to affirm a nonformal and practical rendering of the semantic conception of truth in cases of empirical matters of fact and a nontechnical statement of the wave theory of light but not in cases concerning Darwin's theory of evolution. We cannot draw a general conclusion from inconclusive findings without a statistically significant difference, so comparing the data of question 6 with question 1 would not be very helpful. Nevertheless, we can still say that the content of question 7 having to do with Darwin's selection theory does seem to be pushing people's intuitive responses around when it comes to affirming the synonymy of a general statement with its truth-affirming statement. The difference in the pattern of responses between question 6 and question 7 further suggest that this is the case.

2.5 Third comparison case

All of the above analyses have involved particular instances of the T-Schema where the collected empirical data seem to show that people's intuitive responses regarding synonymy between "p" and "It is true that p" tends to fluctuate according to p's specific content. So sensitive are people's responses that they seem to recognize a distinction between different scientific theories. The question remaining to be answered is whether removing specific content and replacing it with a free variable, namely p, would make a difference for the pattern of responses. Næss hadn't presumed so, but in a reevaluation of Næss's data our hypothesis would be that in fact there would be a different pattern for responses in specific instances of the T-Schema and the generalized form of the T-Schema, using a free variable such as p. In this third comparison of data reported by Næss, we compare empirical matters of fact (Question 1) with something on the order of ES (Question 8).

The question asked of study participants was modified slightly for the iteration of the experiment using a free variable: "Are some of the following kinds of sentences for you expressive of the same assertion, if you replace S by actual instances of sentences of the kind which express assertions (declarative sentences)[?]" (Næss, 1953: 40). And participants received the following list:

A. S
B. S is true.
C. S is perfectly certain.
D. S in (*sic.*) extremely probable. (Næss, 1953: 41)

When we compare responses to Question 8 with responses to Question 1, empirical matters of fact, then we have the following data set (see Table 7.3.).

Table 7.3 Comparison of responses to Q1 and Q8

	Q1 (%)	Q8 (%)
Synonymy affirmation	87	64
Synonymy denial	13	36

There is no doubt that nonphilosophers tended to affirm the synonymy of "*p*" and "It is true that *p*" for both questions, but they tended to shy away from synonymy claims for Q8 more so than for empirical matters of fact in Q1. Just as in the previous two comparisons, the statistical analysis is our own and shows that the difference was statistically significant (χ^2 (1) = 15.74, p < .001).

2.6 A discussion of the third comparison case

Næss writes, "The conspicuously high frequencies of the synonymity affirmation . . . found as answers" to question 1 "are not maintained in the general case inquired about in question 8" (Næss, 1953: 18). When people have not been provided with the content of *p*, they tend not to affirm that "*p*" and "It is true that *p*" are synonymous. Whatever one might say about Næss's data and analysis, it would be impossible to ignore that nonphilosophers harbor an interesting notion of truth that on occasion aligns with Tarski's semantic conception of truth and on others fails to do so.

3 Practical variant of Tarski's theory of truth

Since learning of Næss's early investigations into Tarski's formulation of truth, we returned to the project of trying to capture whether and how a Tarski-style formulation, either in its general form (ES) or considered through various instances of ES might fit into the nonphilosopher's view of truth. We report data accumulated over several years. While Edouard Machery has suggested in personal communication that data have an expiry date and like milk goes sour after a month's time, we reject this assessment. Data never expire. Social scientists regularly employ data gathered over several years to perform longitudinal analyses. While we cannot offer a genuine longitudinal study, nor would one be appropriate to the present question, we believe that reporting early data together with the more recent data is a more honest approach and important given the descriptive task of our studies.

3.1 The 2010 and 2014 study

The Yale Experiment Month study conducted in 2010 surveyed 355 people from diverse regions of the world who ranged in age from 18 to 65 years.[9] Responses were collected using *Qualtrics*, an online survey tool, and participants were recruited by means of online announcements and through Amazon's Mechanical Turk ("M-Turk"). M-Turk participants were compensated approximately $1.25 for 10 to 15 minutes of their time. Participants were randomly presented with one of nine short vignettes.[10] For each group, we employed a within-subject study design. All participants were then asked to respond to sixteen statements about truth using a 5-point Likert scale, from 1 ("strongly disagree") to 5 ("strongly agree").[11] Demographic and educational attainment data were also collected. We hypothesized that agreement or disagreement with a practical variant of Tarski's T-Schema would depend upon the condition presented to the respondent. One statement was designed to assess whether an expression of the form "p is true" was equivalent to an expression of the form "p."

The specific statement was:

> [C] When I say that a claim is true, I have said nothing more than what the claim itself expresses.

ANOVA analysis failed to show that exposure to any of the nine conditions had a significant effect on mean participant responses; results reported here aggregate responses from all nine conditions. Unlike our other experiments, these data failed to show that participant's responses to [C] fluctuated according to the prompt they read prior to judging whether they agreed or disagreed with [C] (Barnard and Ulatowski, 2013). Likewise, we did not find any evidence of individual differences in the data having to do with [C] (see Table 7.4) (cf., Barnard and Ulatowski, 2013, ms).

In 2014, a follow-on study ($N = 90$) was conducted in which the same item was presented to a diverse population of study participants. Just as with the 2010 study, responses were collected using *Qualtrics* and participants were recruited by means of online announcements and through M-Turk who were compensated approximately $1.25 for 10–15 minutes of their time. However, in this study, no vignette preceded [C] (see Table 7.5).

Table 7.4 Agreement or disagreement with [C] (Condition)

Strongly agree	Agree	Neither agree nor disagree	Disagree	Strongly disagree
20.8%	30.1%	20.0%	18.6%	10.4%
Expressing agreement (SA/A)		Expressing non-agreement (NAD/D/SD)		
50.9%		49.1%		

Table 7.5 Agreement or disagreement with [C] (No condition)

Strongly agree	Agree	Neither agree nor disagree	Disagree	Strongly disagree
12.2%	38.9%	33.3%	8.9%	6.7%
Expressing agreement (SA/A)		Expressing non-agreement (NAD/D/SD)		
51.1%		48.9%		

Neither the 2010 nor the 2014 empirical studies we conducted showed that there were statistically significant variations in people's responses to a variant of Tarski's semantic conception of truth, as we had hypothesized.[12]

3.2 Discussion of the 2010 and 2014 studies

The aim of these two studies was to determine the degree to which people embrace a deflationary notion of truth and whether that agreement or disagreement was informed by conditions respondents had read prior to giving a response to [C]. We believed that a short vignette preceding [C] would contribute to respondents' agreement or disagreement with [C], thus suggesting that background contextual features influence people's intuitive responses either toward or away from agreement with [C]. These studies, however, were inconclusive, since our hypothesis was not borne out.

What we found was evidence of underlying stability in how people think about truth. So, we decided to return to Næss's 1953 studies and use them as a model for a new experiment more directly focused upon Tarski's theory of truth.

4 Næss redux experiment

For this experiment, participants were recruited through M-Turk, and each respondent was compensated approximately $1.00 for no more than ten minutes of their time.[13,14] Our project attracted a geographically and demographically diverse group of 300 total participants aged 18–65 years. Of these, 183 (60.8 percent) self-identified as male and 118 (39.2 percent) self-identified as female. Further, 160 (53.2 percent) of the participants self-reported educational attainment of a completed undergraduate degree or more. For the first Næss redux experiment, we employed a between-subjects study design where study participants were randomly assigned to one of four conditions modeled on those from Næss's original synonymy study and asked whether they believed the two statements were equivalent or different. Detailed demographic information was also collected.

For the Næss redux experiment, we employed a between-subjects study design, using Næss's original questions as a template for our own slightly modified version. Study participants were randomly selected to receive one of four statements modeled on those from Næss's original synonymy study and asked whether they believed the two statements were equivalent or different. The responses for each group were then compared.

On the working assumption that Næss's assessment was correct, we hypothesized that nonphilosophers' notions on a practical version of Tarski's semantic conception of truth vary according to the content of p. Just as Næss had found that people's intuitive responses to examples of empirical matters of fact differ from propositions that include content about Darwin's evolutionary theory, we believed that we would find a similar pattern of responses.

Næss began by providing participants with criteria for assessing synonymy. Like Næss, we provided respondents with criteria for how they should judge whether two statements are equivalent or different. For each question, participants received the following set of instructions:[15]

> For the following question, you are asked to determine whether certain sentences say the same thing. It is generally assumed that different sentences may occasionally mean the same in the sense that in all or some situations the two sentences express the same thing.
>
> We invite you to use the following criteria:
>
> A sentence is for you different from another sentence if you can imagine possible (but perhaps not actual) circumstances where the first sentence is correct but the second sentence is incorrect, or vice versa.
>
> A sentence is for you equivalent to another sentence if under all circumstances either both sentences are correct or both sentences are incorrect.

Then, study participants were given one of the following statements:

> Q2. There is at least one copy of the Bible in the University Library.
> Q3. It will be raining in Paris on 17 September 2018.
> Q4. There is an ether that oscillates in accordance with the wave theory of light or closely related theories.
> Q5. Darwin's selection theory explains the evolution of populations through change in heritable traits over time.

Table 7.6 Attempted replication of Næss's 1953 synonymy experiment

Q	N	Equivalent (%)	Different (%)
Q2	86	94.2	5.8
Q3	78	85.9	14.1
Q4	67	88.1	11.9
Q5	68	80.9	19.1

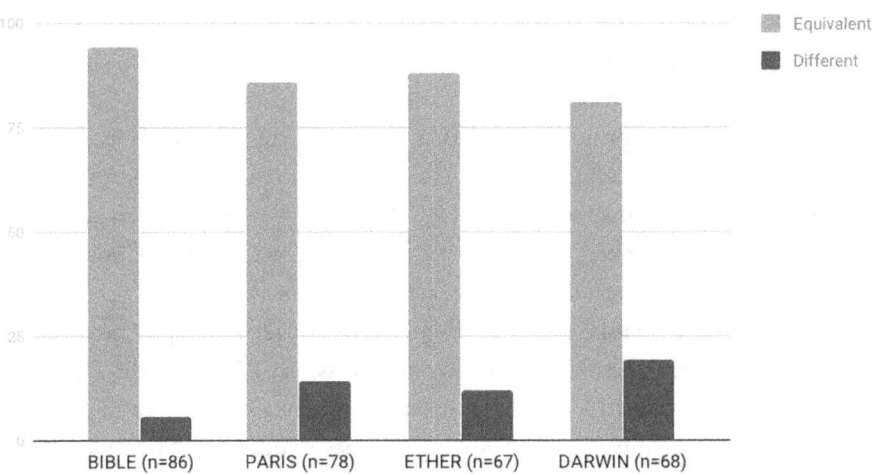

Figure 7.1 Attempted replication of Næss's 1953 synonymy experiment.

Q2–Q5 represented statement A for each study participant. B is one of the statements prefixed with "It is true that . . . " For Q2, A is, for example, "There is at least one copy of the Bible in the University Library" and B is: "It is true that there is at least one copy of the Bible in the University Library." Respondents were asked:

Do sentences A and B express the same thing, or not?

Interestingly, Næss's main results replicated after more than sixty years (see Table 7.6 and Figure 7.1)! The data show a significant difference between people's responses to the Bible case, an empirical matter of fact, and the Darwin case, something on the order of a controversial scientific theory if we take the (mistaken) view that theories are just unconfirmed hypotheses (χ^2 (1) = 2.65112, $p < .011$, $N = 154$). It should be noted that the Darwin data in our set of results wasn't nearly as dramatic a difference as what Næss had found, but there is still a statistically significant difference. Nevertheless, we found variation in people's intuitive responses to expressions of empirical matters of fact and scientific theories.

5 Modified Næss experiment

We completed a second experiment, but this one was a within-subjects design and meant to extend the reliability of Næss's original empirical results. Three hundred participants were recruited online and through M-Turk, and each respondent was compensated approximately $1.00 for no more than 10 minutes of their time.[16] Our project attracted a geographically and demographically diverse group of 300 total participants aged 18–65 years. Of these, 183 (60.8 percent) self-identified as male and 118 (39.2 percent) self-identified as female. Further, 160 (53.2 percent) of the participants self-reported educational attainment of an undergraduate degree or more.

Respondents were asked a series of questions concerning a practical variant of Tarski's T-Schema, one of which was an expression of Convention-T. A set of ten questions concerned whether two statements, for example, p and p is true, were equivalent or different.[17] Following on these questions, each person was asked to respond to a series of demographic questions regarding, for example, their political and religious affiliation, gender, age, education level attained, and socioeconomic status.[18]

For this experiment, we employed a "within-subjects" study design. Study participants received eight statements:

Q6. Even numbers are divisible by two without remainder.
Q7. Every even number is the sum of two prime numbers.
Q8. Unicorns tend to sleep underneath oak trees.
Q9. Emma is honest.
Q10. William Shakespeare's Romeo and Juliet have matching tattoos.
Q11. Dr. Watson and Sherlock Holmes often share a pipe at afternoon tea.
Q12. The old man's liver deftly legislated.
Q13. p

Q6–Q13 represented statement A for each study participant. B is one of the statements prefixed with "It is true that . . . " Q8A is, for example, "Unicorns tend to sleep underneath oak trees" and Q8B is "It is true that unicorns tend to sleep underneath oak trees."

For Q6–Q12, participants received the following set of instructions:

> For the following question, you are asked to determine whether certain sentences say the same thing. It is generally assumed that different sentences may occasionally mean the same in the sense that in all or some situations the two sentences express the same thing.
>
> We invite you to use the following criteria:
>
> A sentence is for you different from another sentence if you can imagine possible (but perhaps not actual) circumstances where the first sentence is correct but the second sentence is incorrect, or vice versa.
>
> A sentence is for you equivalent to another sentence if under all circumstances either both sentences are correct or both sentences are incorrect.

The instructions had to be modified slightly for Q13 (modifications have been italicized).

> For the following question, you are asked to determine whether certain sentences say the same thing. *Assume that we can replace "p" with any sentence.* It is generally assumed that different sentences may occasionally mean the same in the sense that in all or some situations the two sentences express the same thing.
>
> We invite you to use the following criteria:
>
> A sentence is for you different from another sentence if you can imagine possible (but perhaps not actual) circumstances where the first sentence is correct but the second sentence is incorrect, or vice versa.
>
> A sentence is for you equivalent to another sentence if under all circumstances either both sentences are correct or both sentences are incorrect.

Participants were then asked whether they believed QxA and QxB were equivalent or different (see Table 7.7 and Figure 7.2).

The results of Q6–Q12 conform with what we found in the redux experiment reported above. The exception was the Q13 result. In this case, a majority of participants judged that "p" and "It is true that p" were different. The result for Q13 was compared with the responses for the seven other examples and a significant difference was found in each case (see Table 7.6). The results indicate that while people readily agree that instances of "It is true that p" and p are equivalent when content stands in for p, they say that the two are different when no content stands in for p.

Table 7.7 Judgments of equivalence and difference among various statements and their alethically modified counterparts

Q	N	Equivalent (%)	Different (%)	Difference from P
Q6	301	84.4	15.6	$\chi^2(1) = 198.858$, $p < .00001$
Q7	302	87.1	12.9	$\chi^2(1) = 220.139$, $p < .00001$
Q8	302	76.2	23.8	$\chi^2(1) = 144.051$, $p < .00001$
Q9	302	85.4	14.6	$\chi^2(1) = 207.142$, $p < .00001$
Q10	302	83.8	16.2	$\chi^2(1) = 194.713$, $p < .00001$
Q11	303	85.1	14.9	$\chi^2(1) = 205.375$, $p < .00001$
Q12	301	81.7	18.3	$\chi^2(1) = 179.752$, $p < .00001$
Q13	303	27.4	72.6	n/a

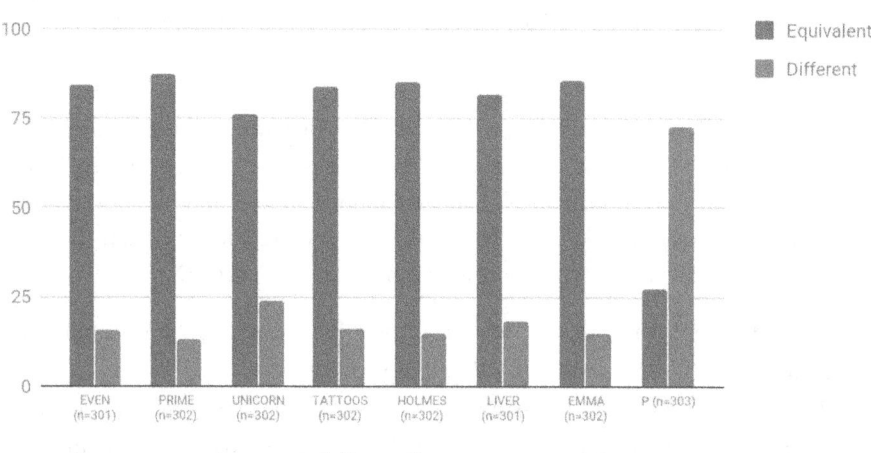

Figure 7.2 Judgments of equivalence and difference among various statements and their alethically modified counterparts.

6 Discussion

While neither we nor Næss can offer a specific theory that can explain the results we report, we think a few preliminary observations are relevant. Let's begin with three meager observations of what the data tell us about empirical studies of philosophical theories of truth. First and most importantly, the "unsurprising" result noted by Tarski (1944: 360) (but due to Næss) that 90 percent of people agree with "a sentence such as 'it is snowing' is true if, and only if, it is snowing" aligns with our data. We believe that our data represent a successful replication of one of the first empirical studies undertaken with the intention of addressing a philosophical problem.[19] If nothing else becomes of the data we report here, it is that Næss's early work, and the results he reported, have been partially vindicated and partially defended from the charge that (a) his work is uninteresting or (b) his experimental design was unsophisticated and wouldn't withstand a more thorough contemporary analysis.

Second, judgments about synonymy are semantic or conceptual, not syntactic. We note that people tended to affirm the equivalence of a variety of instances where "p" and "It is true that p" was replaced by empirical matters of fact, examples of fiction, mathematical examples, and even nonsense. This is good news for some, since such equivalence judgments support views that try to understand truth through Tarski-style biconditionals. The nonsense case, "The old man's liver deftly legislated," is an interesting one because it shows that nonphilosophers can, on occasion, bypass the content of expressions in judgments about the equivalence of "p" and "It is true that p." Free variable cases, such as in the case of Q13, lack specific content, so participants may have been reluctant to judge whether "p" and "It is true that p" are equivalent if they thought that p could be replaced with two different expressions, or that p failed to express any semantically evaluable content. In the case of testing free variables, there are no actual meanings available for language users to compare. The uninterpreted "p" is merely a placeholder for some expression. So, nonphilosophers might not know exactly what to do in such cases and choose to err by suggesting that the two are different and not equivalent. The present data do not allow for a meaningful disambiguation here, so we will have to leave that for a follow-on empirical study.

The inference from our results to conclusions about the acceptability of Tarski-style biconditionals deserves a clear articulation. Some will no doubt note that we did not ask the question: do you accept that "'p' is true if and only if p"? Such an approach would have presumed that ordinary folk understood quotational devices and the truth functional assessment of the biconditional. Our decision to follow Næss and approach the issue through an assessment of synonymy was rooted in a particular understanding of the biconditional. If we begin with a biconditional such as "The figure has ten sides if and only if it is a decagon," the truth of the biconditional depends upon "The figure has ten sides" and "The figure is a decagon" meaning the same thing (and hence having the same truth conditions). So, we asked if two expressions had equivalent or different meanings. We assume that if participants judge that the expressions are equivalent, then this supports the affirmation of the corresponding biconditional. We do not assume that people actually draw that conclusion; rather, if they judge that the expressions are different, we expect that they reject the biconditional. In other words, affirming the equivalence of A and B is consistent with (A iff B), but affirming the difference between A and B entails that one not accept (A iff B). This is an indirect way of approaching the matter, but so long as one maintains a semantically driven reading of Tarski-style biconditionals, it seems reasonable.

Third, and more broadly, we think it is worth considering that the results we report might fit into a larger phenomenon. There is a familiar pattern of divergent judgments when we compare general formulations of moral, logical, and psychological principles to particular cases or instances of these principles in other areas of philosophy and psychology. Tamar Gendler (2007) discusses this phenomenon at length. She suggests that one role played by thought experiments in philosophy is to provide an opportunity to think about the relationship between general formulations of philosophical principles and hypothetical instances of those principles. Further, she notes, it is commonplace to find that judgments about instantiated hypothetical cases often differ from judgments about non-instantiated cases. This is evident in the psychology of reasoning with respect to, for example, the Wason selection task where people's reasoning patterns tend to conform better to *modus ponens* and *modus tollens* in instantiated cases involving, for example, cheating detection. A similar result is also found in a number of familiar thought experiments in ethics, for example, various formulations of the trolley cases. The upshot of these cases, considered collectively, is that we should not be surprised to see divergence between responses to particular instances and general uninstantiated instances of principles such as Tarski's biconditional.[20]

7 Concluding discussion

Our explanation of the data we report here ought to have some implications for deflationary accounts of truth. For the remainder of the chapter, we would like to focus on Tarski's semantic conception of truth (1936/1983, 1944) and Horwich's Minimalism about Truth (1990/1998, 2010).

According to Tarski, we can give a rigorous definition of truth only within a formal language. Formal correctness, especially, requires precision and univocity. Convention-T and instances of ES, give us the content of Tarski's theory. But Tarski also clearly indicated that his theory ought to conform to a pre-theoretic understanding of truth that is expressible in natural language and that this conformity might be supported or undermined by appeal to how ordinary people might employ the notion of truth (e.g., through the use of social scientific surveys). So, a test of whether ordinary people understand and assent to instances of ES (albeit expressed in nonformal language), is a means to better understand the degree to which Tarski achieves this conformity.

The extension of Tarski's formal semantic conception of truth is given by the axioms derived from ES, which collectively yield an inductive definition of truth, specifying all the axioms that serve to specify the extension of the concept of truth. Tarski defined the class of axioms recursively because he understood that no single instance or finite set of instances of ES would suffice to give a general definition of truth. For example, we could imagine an innumerable number of ES sentences:

> "Rusty is Nell's lover" is true if and only if Rusty is Nell's lover.
> "Shakespeare wrote *Hamlet*" is true if and only if Shakespeare wrote *Hamlet*.
> "Duchesse de Bourgogne is a Flanders red ale" is true if and only if Duchesse de Bourgogne is a Flanders red ale.

No matter how many sentences we imagine using ES they would be insufficient to generalize to a formal rendering of a theory of truth. Nevertheless, instances of ES can serve to provide a partial definition of truth that is expressible in ordinary language and thereby available to ordinary language users.

Our data seem to support that the content of instances of ES is easy to understand for nonphilosophers. Not only that, our data show that nonphilosophers tend to believe that affixing a truth predicate or prefix to instances of ES that includes content about empirical matters of fact (e.g., It is true that "snow is white"), mathematical facts (e.g., "3 is a prime number" is true), and even facts about fiction (e.g., It is true that "Sherlock Holmes lives on Baker Street") are equivalent to the instances alone (e.g., snow is white, 3 is a prime number, Sherlock Holmes lives on Baker Street). The data suggest, then, that people understand ES biconditionals. But people tend to believe that the schematic version p and "p" is true are not equivalent, that is, it appears that people can imagine a case where p is correct to use but "p" is true is not correct (or vice versa). It follows that these data have nothing to say about the formal aspects of Tarski's semantic conception of truth. But, if we want to draw any practical lessons from Tarski's theory about how ordinary people employ the notion of truth, we need to account for the differences in how people think about instances of ES as opposed to ES itself.

That leads us to consider how our empirical data further shape the empirical adequacy of a deflationary account of truth that takes as its starting point instances of ES, namely Horwich's minimalist theory about truth ("MT"). According to Horwich, to understand truth is to comprehend how ES works, which is captured by three core theses. First, the propositions expressed by non-paradoxical instances of ES make up the axioms of Horwich's MT. Second, the inferential role of "true" as a logical predicate is not purely redundant. Finally, since there isn't an uncontroversial way of generalizing from instances of ES, Horwich proposes that we have a disposition to accept all uncontroversial and non-paradoxical sentences instantiating ES as capturing the entirety of MT.

If we accept the three guiding principles of MT, then minimalism is explanatorily, conceptually, logically, and epistemologically fundamental. It is explanatorily fundamental because everything we are able to do with the truth predicate can be explained, according to Horwich, in terms of instances of ES. Second, MT is said to be conceptually fundamental since no substantive concept underwrites instances of ES; such instances are brute.[21] Third, it is natural for the truth predicate to enable the explicit formulation of schematic generalizations, and, therefore, allow MT to be logically fundamental. Finally, according to minimalism, instances of ES needn't be justified by anything more immediately known or anything more obvious to us than non-paradoxical cases of ES; MT is epistemologically fundamental.

Our empirical data speak to the epistemological fundamentality of Horwich's MT and show that the axioms of MT seem to represent an intuitive position among nonphilosophers. Since nonphilosophers tend to accept instances of ES, it seems clear that MT is fundamental. Given the acceptability of instances of ES, there is reason to believe that Horwich's MT might be descriptively adequate. MT, or at least a practical variant of it, is among those views upheld by nonphilosophers.[22] It is an important and empirically testable consequence of Horwich's minimalist conception of truth that ordinary people would, in fact, manifest the disposition to accept instances of ES. The data that we have collected so far are consistent with the hypothesis that people do manifest that disposition.

Tarski's distillation of a rigorous account of truth has had a lasting impact on philosophical logic. Despite this, or perhaps because of it, in later work Tarski considered empirical studies on the nonphilosopher's notion of truth to be an informative means of coming to appreciate the nuances of the workings of truth in natural language (cf. Barnard and Ulatowski, 2016, Ulatowski, 2017). In a series of studies completed from the 1930s to the 1960s, Arne Næss collected and analyzed intuitive responses from nonphilosophers to questions concerning truth, synonymy, certainty, and probability. Among the formulations of truth studied by Næss was ES. Our study has called attention not only to Næss's early findings but to a series of experimental results we've collected that suggest people respond affirmatively to the synonymy of a statement and its alethically modified counterpart when that statement has content, but people are reluctant to affirm instances of ES when it is more abstract and includes free variables. So, a response to our titular question: does anyone really think that ⌜φ⌝ is true if and only if φ? is a qualified one, insofar as our data point out that nonphilosophers tend to think that ⌜φ⌝ is true if and only if φ when we replace φ with some content but tend to think otherwise when φ is not replaced by some semantic content.

Notes

1 We adopt quasi-quotation marks, a phenomenon that originated with Quine (1940), as a linguistic device to indicate that φ *stands for* linguistic expressions and is *used as* that linguistic expression in a different instance. In other words, φ is an expression, and ⌜φ⌝ is a name of that expression. However, it is important to note that our doing so does not commit us to ⌜φ⌝ standing for propositions, sentences, assertions, utterances, or other technical terms of philosophical art. We wish to avoid any theoretical discussion of these technical details.
2 This is sometimes called the T-Schema, but, at least for this chapter, we would like to abbreviate it as ES.

3   For a more detailed discussion of how compositionality is connected with the material adequacy condition, we recommend that the reader refer to Kirkham (1992), chapter 5. Not all deflationists about truth, of course, need to adopt a compositional semantics, for example, Paul Horwich (1990/1998).
4   Here we sacrifice precision for clarity. More precisely the second instance of the variable is really a second order sign that names the first variable. What is interesting here is that the use of the "It is true that . . ." device does not interfere with the naming-named relation.
5   Besides Carnap, Næss also credits "Professor D. Rynin and F. Fluge" for providing "valuable suggestions and help" leading up to the publication of the 1953 study.
6   We raise the issue here about the criteria because in an anonymous referee report received on a different project in which one of us (Ulatowski) provided a summary of Næss's 1953 studies, the referee worried "God knows how a study participant may have understood the directions?" We believe that this is an unusual comment about Næss's work because he paid meticulous attention to experimental design. Before running his experiments on study participants, Næss would run a pilot study to ensure that respondents were understanding the directions that would eventually appear in the experiment.
7   E through G were not asked in all iterations of the questionnaire. For this reason, we have separated them out from A through D using an asterisk.
8   An audience comment responding to Ulatowski on Næss-related material suggested that Næss's methodology was flawed for Question 7 because it appears that he didn't provide study participants with a formulation of Darwin's theory. Næss employed a generic, practical formulation of Darwin's evolutionary theory that wasn't necessarily consistent with the latest information associated with evolutionary theory. His calling it a "formulation of Darwin's selection theory" is a means of relaying to a sophisticated reader that he understands there is more to the debate than a quip outlined in so few characters.
9   Our study received approval from the UNLV IRB certified in a letter.
10 For example, the "Anna" vignette was: "Anna has performed a simple calculation and discovered that 30 + 55 = 85"; the "Charles" vignette was: "Charles reads a lot. Charles has discovered the claim that aluminum is made by refining bauxite. He has never refined bauxite, nor is he an expert in metals. Charles thinks that aluminum is made by refining bauxite." A full list of these vignettes is included as an appendix to Barnard and Ulatowski (forthcoming).
11 Independent of the sixteen statements, Question #1 was specifically associated with the condition and used to test whether the subjects were giving random responses.
12 If one aggregates the responses as we have with only clear affirmative responses indicating agreement, then there is no apparent difference between the percentage

who agree with [C] and those who disagree. If one ignores those who chose the midpoint response, then for those who express a clear position the results indicate overall agreement with [C] (76.7percent agree or strongly agree in 2010; 63.7percent agree or strongly agree in 2014).

13 Responses were collected through the University of Waikato's *Qualtrics* internet portal. Co-investigators sought and received IRB approval through the University of Waikato and University of Mississippi (Approval #18x-048).

14 M-Turk is populated by people who have access to the internet and who are generally computer literate. We can understand why some might resist the use of this respondent pool, but we foresaw no reason not to employ the M-Turk population. Beyond that, we placed few restrictions on those who could respond to the survey questionnaire. Only respondents who self-identified as being a part of sensitive populations, for example, children under the age of eighteen, prisoners, etc., were forbidden from submitting responses. Published studies of the M-Turk population suggest that the sample is in many ways superior to the standard social psychological sample of college students (Buhrmester et al., 2011; Paolacci et al., 2010).

15 The instructions were vetted by a pilot study group of nonphilosophers to check for comprehension of the criteria prior to conducting the experiment.

16 Responses were collected through the University of Waikato's *Qualtrics* internet portal. Co-investigators sought and received IRB approval through the University of Waikato and University of Mississippi (Approval #18x-048).

17 There were three questions asking respondents for feedback concerning the role evidence plays in understanding truth, but those are set for inclusion in another paper. See Barnard and Ulatowski (ms).

18 No other personal information, such as name, address, email address, or phone number, was collected from respondents in order to preserve their anonymity.

19 Of course, our claim isn't that Næss's empirical studies were the first presaging contemporary experimental philosophy. That title might belong to Fraser (1891, 1893). And concern for the role intuitions play in philosophical argument goes as far back as the first issues of the journal *Mind* (cf. Thomson, 1878a, 1878b; Davidson, 1882).

20 See Gendler (2007) for extended discussion.

21 Instances of ES do not follow from definitional relations holding between the concept of truth and more basic concepts in terms of which "true" can be defined. Instances of TS are analytic, in addition to being necessary and *a priori*.

22 We realize that Anil Gupta's generalization problem (1993) attempted to address the epistemological dimension of Horwich's MT. While we would like to show how our empirical data reported in this chapter speaks to Gupta's generalization problem, space constraints preclude an adequate discussion here.

References

Alexander, J. (2012). *Experimental Philosophy*. New York, NY: Polity Press.

Barnard, R. and Ulatowski, J. (2013). Truth, correspondence, and gender. *Review of Philosophy and Psychology*, 4(4):621–38.

Barnard, R. and Ulatowski, J. (2016). Tarski's 1944 polemical remarks and Næss' "experimental philosophy". *Erkenntnis*, 81(3):457–77.

Barnard, R. and J. Ulatowski (forthcoming). Objectivity, a core truism? *Synthese*, http://doi.org/10.1007/s11229-017-1605-7.

Barnard, R. and J. Ulatowski. (ms). Just the facts: Evidence, gender and the ordinary view of truth. University of Waikato, 30pp.

Buhrmester, M., Kwang, T. and Gosling, S. (2011). Amazon's Mechanical Turk: A new source of inexpensive, yet high quality, data? *Perspectives on Psychological Science*, 6(1):3–5.

Chapman, S. (2018). The experimental and the empirical: Arne Næss' statistical approach to philosophy. *British Journal for the History of Philosophy*, 26(5):961–81.

Davidson, W. (1882). Definition of Intuition. *Mind* (Old Series), 7(26):304–10.

Fraser, A. (1891). Visualisation as a chief source of the psychology of Hobbes, Locke, Berkeley, and Hume. *The American Journal of Psychology*, 4(2):230–47.

Fraser, A. (1893). The psychological basis of Hegelism. *The American Journal of Psychology*, 5(4):472–95.

Frege, G. (1919/1956). The thought. *Mind*, 65(259):289–311.

Gendler, T. (2007). Philosophical thought experiments, intuitions, and cognitive equilibrium. *Midwest Studies in Philosophy*, 31:68–89.

Gupta, A. (1993). A critique of deflationism. *Philosophical Topics*, 21(2):57–81.

Horwich, P. (1990/1998). *Truth*, Revised edition. Oxford: Clarendon Press.

Horwich, P. (2010). *Truth, Meaning, Reality*. Oxford: Oxford University Press.

Kirkham, R. (1992). *Theories of Truth: A Critical Introduction*. Cambridge, MA: MIT Press.

Knobe, J. and Nichols, S. (Eds.) (2008). *Experimental Philosophy, Volume 1*. Oxford: Oxford University Press.

Knobe, J. and Nichols, S. (Eds.) (2014). *Experimental Philosophy, Volume 2*. Oxford: Oxford University Press.

Næss, A. (1938a). *"Truth" as Conceived by Those Who Are Not Professional Philosophers*. Oslo: I Kommisjon Hos Jacob Dybwad.

Næss, A. (1938b). Commonsense and truth. *Theoria*, 4(1):39–58.

Næss, A. (1953). *An Empirical Study of the Expressions "True," "Perfectly Certain," and "Extremely Probable, Avhandlinger Utgitt av Det Norske Videnskaps-Akademi I Oslo II. Hist.-Filos. Klasse 1953. No. 4*. Oslo: I Kommisjon Hos Jacob Dybwad.

Quine, W. V. O. (1940). *Mathematical Logic*, Revised edition. Cambridge, MA: Harvard University Press.

Paolacci, G., Chandler, J. and Ipierotis, P. G. (2010). Running experiments on Amazon Mechanical Turk. *Judgment and Decision Making*, 5(5):411–19.

Ramsey, F. (1927). Facts and propositions. *Proceedings of the Aristotelian Society*, 7(Supplementary):153–70.

Sytsma, J. and Livengood, J. (2015). *The Theory and Practice of Experimental Philosophy*, Toronto: Broadview Press.

Tarski, A. (1944). The semantic conception of truth and the foundation of semantics. *Philosophy and Phenomenological Research*, 4(3):341–76.

Tarski, A. (1969). Truth and proof. *Scientific American*, 220(6):63–77.

Tarski, A. (1936/1983). The concept of truth in formalised languages. In J. Woodger (Ed.), *Logic, Semantics, Metamathematics*, 152–278, Indianapolis: Hackett.

Thomson, D. G. (1878a). Intuition and inference. *Mind* (Old Series), 3(11):339–49.

Thomson, D. G. (1878b). Intuition and inference. *Mind* (Old Series), 3(12):468–79.

Ulatowski, J. (2016). Ordinary truth in Tarski and Næss. In A. Kuzniar and J. Odrowąż-Sypniewska (Eds.), *Uncovering Facts and Values*, 67–90. Leiden, The Netherlands: Brill.

Ulatowski, J. (2017). *Commonsense Pluralism about Truth: An Empirical Defence*. Basingstoke: Palgrave Macmillan.

8

New Foundations for Fuzzy Set Theory

Igor Douven[1]

1 Introduction

Chances are your mathematics teacher introduced set theory by asking you to abstract from some kind of container (a crate, or a box, or a bag, etc.) with a number of objects (bottles, eggs, apples, etc.) in it. "Think of the crate as the set, and of the bottles as the elements—it's just that in set theory we don't care about what the crate looks like, or what it is made of, nor do we care about how the things that are in it are arranged." For many students, the analogy between sets and crates, and between a set's elements and concrete objects located in some container, is helpful; in the very first stage of learning set theory, when students become familiar with its basic concepts, the analogy provides a simple yet powerful image the student can resort to.

Soon, when the student is introduced to the elementary set-theoretic operations, the container-with-contents image may no longer be so useful—it is difficult to imagine eggs that are in one box to be simultaneously in another box—and may have to be replaced by a team-and-members image: members of the school's football team can very well also be members of the school's basketball team. The team-and-members image may carry the student a long way, certainly if she is able to abstract from the fact that teams tend to have only finitely many members.

In standard set theory, membership is categorical: an item either is or is not an element of a given set; there is nothing in between. This feature of the theory is strongly suggested by both the container-with-contents image and the team-and-members image. An egg balancing on the edge of a box simply is not in the box, and someone soon to join a team is not a member yet; similarly, someone soon to leave the team is, at the moment, still a member.

While helpful, these images are also somewhat misleading. Set theory is supposed to be general, but it is not difficult to conceive of cases of membership—of sorts—that are quite unlike the eggs-in-a-box or member-of-a-team type. For example, there are objects of which it seems wrong to say, without any qualification, that they are red, but equally wrong to say, without any qualification, that they are *not* red. Such objects, we would like to say, are to some—but only to some—extent members of the set of red things. Or consider a man of height 1.79 meter. That man is neither clearly a member of the set of tall men nor clearly outside that set. It seems that he belongs to the set to some degree—just not to the fullest degree. As a result, the membership relation from standard set theory, commonly symbolized as $\in$, is not so useful here.

This led Zadeh (1965) to propose a new kind of set theory—fuzzy set theory—whose hallmark feature is a *graded*, or *fuzzy*, membership relation. Formally, where Ω is a class of objects (for which we will use the variable ω), a fuzzy set $\tilde{S}$ in Ω is defined as a (classical) set of ordered pairs, $\{\langle \omega, \mu_{\tilde{S}}(\omega)\rangle \mid \omega \in \Omega\}$. Here, $\mu_{\tilde{S}}(\cdot)$ is a function from Ω into the $[0,1]$ interval, which indicates, for each $\omega \in \Omega$, the degree to which ω belongs to $\tilde{S}$. The closer $\mu_{\tilde{S}}(\omega)$ is to 1, the "more" ω belongs to $\tilde{S}$, with the extremes 0 and 1 indicating full non-membership and full membership, respectively. Classical or ordinary sets (or crisp sets, in more recent terminology) are limiting cases of fuzzy sets, inasmuch as the membership function for such sets takes values only in $\{0,1\}$.

Zadeh (1965) shows how all the key notions of standard set theory have natural extensions in fuzzy set theory. For example, the empty set is defined as the set $\emptyset = \{\langle \omega, \mu_\emptyset(\omega)\rangle \mid \omega \in \Omega\}$ where $\mu_\emptyset(\omega) = 0$ for all $\omega \in \Omega$. And where $\tilde{S} = \{\langle \omega, \mu_{\tilde{S}}(\omega)\rangle \mid \omega \in \Omega\}$, the fuzzy complement of $\tilde{S}$ is defined as $\tilde{S}^c = \{\langle \omega, 1 - \mu_{\tilde{S}}(\omega)\rangle \mid \omega \in \Omega\}$. The other standard set-theoretic operations also have parallel definitions in fuzzy set theory:

$$\tilde{S}_1 \tilde{\subset} \tilde{S}_2 := \mu_{\tilde{S}_1}(\omega) \leq \mu_{\tilde{S}_2}(\omega), \text{ for all } \omega \in \Omega;$$

$$\tilde{S}_1 \tilde{\cup} \tilde{S}_2 := \{\langle \omega, \max[\mu_{\tilde{S}_1}(\omega), \mu_{\tilde{S}_2}(\omega)]\rangle \mid \omega \in \Omega\};$$

$$\tilde{S}_1 \tilde{\cap} \tilde{S}_2 := \{\langle \omega, \min[\mu_{\tilde{S}_1}(\omega), \mu_{\tilde{S}_2}(\omega)]\rangle \mid \omega \in \Omega\}.$$

Zadeh proves that, given these definitions, De Morgan's laws and the distributivity laws, which hold for classical sets, hold for fuzzy sets as well. He also defines the algebraic product and algebraic sum of two sets, which are basically obtained by multiplying and, respectively, adding up the graded membership functions associated with the sets.

Zadeh's 1965 paper was the start of what was to become the highly successful research program of fuzzy mathematics, in which fuzzy set theory served as the basis for developing fuzzy algebra, fuzzy calculus, fuzzy graph theory, fuzzy logic, fuzzy probability theory, and more (see, for example, Dubois and Prade, 1979, 1982). The fruits of this paradigm have found application in a great variety of domains. Everyone knows washing machines "with fuzzy logic," which are typically more expensive than the ones that do your laundry on the basis of (presumably) classical logic.[2] Other applications concern traffic control, heating systems, bio-informatics, decision-making support systems, stock market forecasting, and image enhancement, to name just a few areas where tools from fuzzy mathematics have become popular.

Despite these manifest successes, mathematicians tend to hold the whole research program in low esteem, and you are much more likely to find a course on fuzzy set theory or any other branch of fuzzy mathematics offered by a department of electrical engineering or operations research or computer science than by a mathematics department. Mathematicians' discontent stems from a sense that the notion of graded membership, which the program ultimately rests on, is too subjective and too nebulous to play any serious scientific role; in their view, fuzzy mathematics is built on sand (see, for example, Lindley, 2004).

Starting in the 1970s, vagueness has become a much-debated topic in analytic philosophy. While Zadeh's work is not frequently cited in the vagueness debate, one of the debate's most central questions concerns the metaphysical status of borderline cases, that is, cases falling under a concept only to *some* extent; and the question of what a borderline case is, essentially *is* the question of what (nontrivial) graded membership is, just asked in a slightly different vocabulary.[3] In Zadeh's fuzzy set theory, graded membership figures as a primitive notion. By contrast, philosophers have sought to analyze "borderlineness" in terms of putatively more fundamental and better understood concepts.

Much fine-crafted work has come out of this debate, which however is not to say that we got any closer to a shared view of borderlineness or graded membership. To the contrary, the proposals made have been all over the place, with some authors arguing that the world is inherently vague (e.g., Unger, 1980), others arguing that vagueness is a semantic phenomenon (e.g., Kölbel, 2010), again others that the issue is epistemic (e.g., Williamson, 1994), and still others that vagueness resides primarily in the human mind (e.g., Douven et al., 2013).

Accordingly, different philosophers have held that vagueness is properly studied as a chapter of metaphysics, of the philosophy of language, of epistemology, and of the philosophy of mind, respectively. And philosophers agreeing on which subfield of philosophy the study of vagueness belongs to typically *still* disagree on the finer details of how vagueness is to be accounted for, and specifically also on what borderlineness is.

If what are supposed to be experts in conceptual clarification so dismally fail to reach a consensus on what graded membership amounts to, then a collective endeavor that might have resulted in a vindication of fuzzy set theory actually further discredits it.

But perhaps we have been looking for a vindication in the wrong place. In the following, I suggest that recent work in cognitive psychology holds promise of explicating the notion of graded membership in a way that should leave no doubt as to its scientific respectability. I start by reviewing how Kamp and Partee propose to model vagueness using so-called completions (Section 2).[4] As will be seen, these authors came close to providing an operationalization of graded membership, but they ultimately failed. Later work by Decock and Douven (2014) showed how an operational definition of graded membership flows naturally from Kamp and Partee's semantics if this is embedded in the conceptual spaces framework made popular by Gärdenfors (2000, 2014) and others (Section 4). In the meantime, empirical evidence has accrued supporting the descriptive accuracy of Decock and Douven's proposal (Section 5). That, I believe, is as close as we have come to a foundation for fuzzy set theory, and also as close as we need to get: we have a crystal-clear definition of graded membership in logico-mathematical terms (indeed we have such a definition for each fuzzy set), which implies a measurement procedure, and applications of this procedure have so far led to precisely the results that were predicted on the basis of detailed knowledge about the structure of concepts.

2 Completions and graded membership

Kamp and Partee's proposal assumes prototype theory, according to which concepts tend to have maximally representative instances—so-called prototypes (Rosch, 1973). In the context of prototype theory, someone looking to define a graded membership relation may well be tempted to equate degree of membership with similarity to the prototype. For instance, one might think that the degree to which a given animal is a bird equals the similarity of that animal to the bird prototype. However, for reasons already pointed out in Osherson and Smith (1981), that idea will not work: even though ostriches are less typical birds than (say) robins, they are still birds to the maximal degree.

Many have thought that this simple observation was enough to show that attempts to define graded membership in terms of similarity to prototypes were doomed from the start. But Kamp and Partee were right to point out that the observation only shows that we cannot simply *equate* graded membership with similarity to a prototype, which leaves open the possibility of successfully defining graded membership *in terms of* prototypes and similarity relations. In fact, providing such a definition is exactly what Kamp and Partee (1995) hoped to accomplish. Their aim is not quite to provide a definition of graded membership for all the predicates of English or for those of any other natural language—given the current state of semantics, that would be overly ambitious. Rather, they consider a language with "simple" predicates, such as "round," "yellow," "adult," "man," and others with monolexemic expressions in English. For this language, they aim to define a semantics that can model vagueness—insofar as that applies to the language's predicates—and whose central semantic concept is that of graded membership.

The semantics they propose consists of a domain of discourse, which comprises all items the language allows one to talk about; a partial model, which assigns an extension and an anti-extension to every predicate in the language; and a set of possible completions.

To explain what completions are, notice that, in Kamp and Partee's semantics, the extension and anti-extension of a predicate taken together need not cover everything in the domain of discourse. Rather, in the extension of a predicate we find all items to which the predicate definitely applies, while in its anti-extension we find all items to which the predicate definitely does *not* apply. But there may be items to which the predicate applies *in*determinately, that is, to a certain extent only. A completion assigns some of those indeterminate cases to the extension of the predicate and all others to the anti-extension.

Fine (1975) and other authors had already introduced the idea of using completions to deal with vague predicates, but Kamp and Partee were the first to combine the idea of completions with prototype theory. This allows them to distinguish between completions that do, and those that do *not*, respect orderings with respect to similarity to prototypes. In their semantics, Kamp and Partee only consider the former—so those that, whenever they assign an item x

to the extension of a predicate P, also assign to that extension all items y which are at least as similar to the P-prototype as x is. Suppose, for example, that it is indeterminate whether a given shade s is blue. Then, in Kamp and Partee's semantics, a completion that assigns s to the extension of "blue" will also assign to that extension any shade s' that is bluer than the former. By contrast, there will be completions that assign s', but not s, to the extension of "blue."

If there are only finitely many items in the domain of discourse, then there are also only finitely many completions in the model. In that case, Kamp and Partee's definition of graded membership can be stated simply in terms of proportions. For instance, the degree of blueness of a given item can then be said to equal the proportion of completions that assign that item to the "blue" extension. Note that, on this proposal, every clear instance of blueness will count as blue to the maximum degree (which is 1) while every clear non-instance of blueness will count as blue to the minimum degree (which is 0). That is surely as it should be. As for the indeterminate cases, if (say) 59 percent of the completions assign an item to the extension of "blue," then that item has a degree of blueness of 0.59.

If, as is more realistic to assume, the domain of discourse is infinite, then things get more complicated. The main difference is that in this case the degree to which an item x is a P-member is not the *proportion* of completions that assign x to the extension of P but the (normed) *measure* of those completions, in the sense of mathematical measure theory (in which a measure is basically just the generalization of the ordinary concepts of length, area, and volume to arbitrary numbers of dimensions; see, for example, Halmos, 1974).

While there is nothing complicated to this idea per se, a problem arises from the fact that, with an infinite number of completions in the model, it is not obvious how to obtain a *unique* measure over them. In the finite case, equating degrees of membership with proportions amounts to weighing all completions equally—which also seems to be the only reasonable thing to do here. However, that option is not available in the infinite case, at least not if we want the resulting measure to be countably additive—as Kamp and Partee do, and as also seems a reasonable requirement. Kamp and Partee (1995: 153) acknowledge this problem, adding that it is not clear to them which criteria would allow one to fix a unique graded membership measure.

Decock and Douven (2014) have proposed a solution to the problem that heavily relies on the conceptual spaces framework. Specifically, they argue that the core elements of Kamp and Partee's proposal can all be naturally interpreted in a particular version of the conceptual spaces framework, and that by doing so, the geometry of conceptual spaces provides the tools to fix a unique graded membership function in a natural and straightforward way. To see how the proposal works, at least the basics of the version of the conceptual spaces framework Decock and Douven assume must be known. I first describe these basics and then turn to Decock and Douven's proposal.

3 The conceptual spaces framework

Conceptual spaces are mathematical objects, specifically one- or multidimensional metric spaces whose dimension or dimensions represent fundamental qualities that items may have. Distances in a conceptual space, as measured by the associated metric, are to be interpreted as representing dissimilarities: the greater the distance between two items as represented in the space, the more dissimilar these items are in the respect (e.g., color, taste, smell, ...) the space is meant to model. For instance, the further apart two items are as represented in color space (see below for details), the more dissimilar these items' colors are. These items could still be very similar in other respects; if they are, this will be reflected by the distances between their representations in other conceptual spaces (e.g., one item might be a green apple and the other a red apple; they might still have a very similar taste). Most importantly, conceptual spaces allow us to conceive of concepts geometrically, as *regions* in a mathematical space.[5]

The aim of this chapter is to demonstrate that fuzzy set theory can be given a solid foundation by showing that the fuzzy membership relation is scientifically kosher. If conceptual spaces—on which the proposal is going to rest—were of doubtful standing themselves, perhaps just by being subjective, that would undermine our goal. Therefore, some words need to be said about how we arrive at conceptual spaces.

To construct a conceptual space, we first need to collect some data, which can be of a number of different kinds. The most commonly used data are similarity ratings, confusion probabilities, and correlations. For instance, the shape space to be looked at in Section 7 was constructed on the basis of a large number of similarity ratings: each of more than 1,000 participants was shown a great number of pairs of container-like objects, and the participants were asked to rate how similar to each other the objects were. An alternative approach would have been to show the objects in each pair not simultaneously but shortly after each other, where both would then have appeared for some milliseconds only and participants would have been asked to indicate whether they thought the objects were the same or different. That would have yielded confusion probabilities (probabilities that different stimuli would be judged to be identical), which could equally have served as input for constructing the shape space.[6] And for an example of a conceptual space—specifically, a space for representing conversational and conventional implicatures—constructed on the basis of correlations (in that case, correlations between truth judgments for statements generating implicatures), see Douven (2019a).

Then, at the first stage of construction of the space, the collected data are fed into some statistical dimension-reduction techniques, which turn data into geometric objects. There is a variety of techniques available for this purpose, the one most commonly used being multidimensional scaling; other standard techniques are principal component analysis and nonnegative matrix factorization. What one aims to accomplish at this stage is not just to obtain *some* metrical space. Rather, the goal is to arrive at a space that is low-dimensional (ideally, with no more than three dimensions), that in addition is interpretable, meaning that we can associate each dimension with some fundamental attribute which the items in the target domain (from which the stimuli used to elicit the data were taken) can be said to possess to differing degrees, and finally, that has good fit, meaning that it faithfully represents similarities among those items. Given that adding dimensions to a space always leads to better fit, there is a tradeoff to be made between the first and third desideratum: we want a low-dimensional space that still has acceptable fit.

There is no guarantee that the first stage is successful. But it if is, then we still only have a similarity space, which is yet to be furnished with concepts. There are different ways to turn a similarity space into a conceptual space. On a proposal made popular by Gärdenfors (2000: Chapters 3 and 4), we obtain a conceptual space from a similarity space by locating, within the latter, the prototypes of various concepts, and then using these to generate a so-called Voronoi tessellation of the space.

Prototypes were already mentioned in the previous section. As for the other element in Gärdenfors's proposal, a Voronoi tessellation of a given space partitions that space into disjoint cells in such a way that each cell is generated by a certain point (the "generator point") in the sense that it contains all and only those points in the space that lie no closer to the generator point of any other cell than to its own generator point. Points that lie equally close to two or more generator points make up the boundaries between cells, belonging to each of these cells. To make this formally precise, let S be an m-dimensional space with associated metric δ, and let $P = \langle p_1,\ldots,p_n \rangle$ be a sequence of pairwise distinct points in S. Then the region

$$v(p_i) := \{p \mid \delta(p,p_i) \leq \delta(p,p_j), \text{ for all } j=1,\ldots,n \text{ with } j \neq i\}$$

is the *Voronoi polygon/polyhedron associated with p_i*, and

$$V(P) := \{v(p_i) \mid i \in \{1,\ldots,n\}\}$$

is the *Voronoi tessellation on S generated by P*. For an illustration, see the left panel of Figure 8.1, which shows a Voronoi tessellation on a two-dimensional Euclidean space.

If we let the points representing prototypes in a given similarity space serve as generator points for a Voronoi tessellation of that space, then, as Gärdenfors (2000, chapter 3) argues, we get a partition of the space whose cells can be plausibly interpreted as the concepts associated with the selected prototypes. Specifically, the resulting partition has several features that are natural desiderata for a conceptual system, most notably that the cells are all convex.[7]

Decock and Douven assume an amended version of Gärdenfors's framework, which was proposed in Douven et al. (2013) in an attempt to do away with

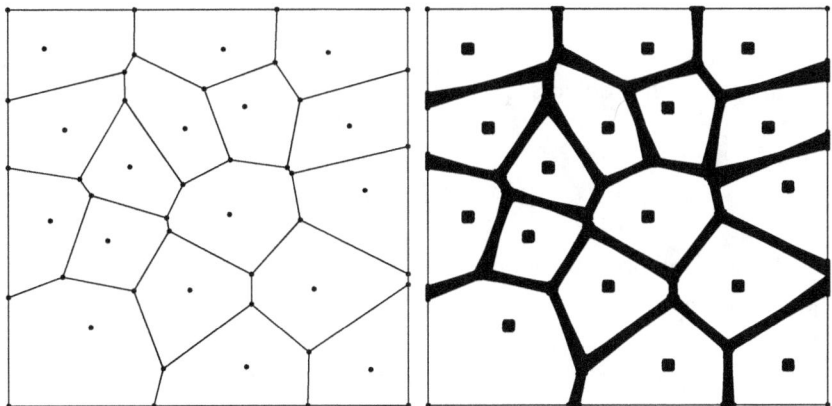

Figure 8.1 The left panel shows one of the ordinary Voronoi tessellations that make up the collated Voronoi tessellation shown in the right panel.

some idealizations of that framework. Most notably, Gärdenfors assumes that each concept has exactly one prototype, an assumption that is known not to be entirely realistic. This is already introspectively clear, for we have no difficulty imagining more than one shade of red we would designate as being *typically* red. But the assumption of unique prototypes also conflicts with experimental data; see, for instance, Berlin and Kay (1969), Douven (2016), and Douven et al. (2017). Thus, Douven et al. (2013) seem right to claim that it is more realistic to equip similarity spaces with prototypical *regions* rather than just with prototypical *points*.

Once we admit prototypical regions, however, we must consider what remains of the technique of Voronoi tessellations, which assumes generator points and not generator regions. Douven et al. (2013) propose replacing the Voronoi tessellations as introduced above by what they call "collated Voronoi tessellations," which are created by (what one could think of as) superimposing ordinary Voronoi tessellations. Informally, to generate a collated Voronoi tessellation on a space, we consider all possible sequences of points from that space which contain exactly one point from each prototypical region. To each such sequence corresponds an ordinary Voronoi tessellation, and we get a collated Voronoi tessellation simply by overlaying them all over each other. See the right panel of Figure 8.1 for an example of a collated Voronoi tessellation.

To make this more formal, let $R = \{r_1,\ldots,r_n\}$ be a set of disjoint prototypical areas in a given space S with metric δ. Then we define

$$\Pi(R) := \prod_{i=1}^{n} r_i = \left\{\langle p_1,\ldots,p_n\rangle \,|\, \text{for all } i \in \{1,\ldots,n\}, p_i \in r_i\right\}$$

to be the set of all ordered sequences $\vec{p} = \langle p_1,\ldots,p_n\rangle$ such that $p_i \in r_i \in R$, for $1 \leq i \leq n$, and

$$\mathcal{V}(R) := \{V(\vec{p}) \,|\, \vec{p} \in \Pi(R)\}$$

to be the set of all ordinary Voronoi tessellations on S generated by elements of $\Pi(R)$. Furthermore,

$$\underline{v}(p_i) := \left\{p \,|\, \delta(p, p_i) < \delta(p, p_j), \text{ for } j = 1,\ldots,n, \text{ with } j \neq i\right\}$$

is the *restricted Voronoi polygon/polyhedron associated with* p_i, and the region

$$\underline{u}(r_i) := \bigcap \{\upsilon(p) \mid p \in r_i \land \upsilon(p) \in V(\vec{p}) \in \mathcal{V}(R)\}$$

is the *collated Voronoi polygon/polyhedron associated with* r_i. Finally, the set

$$\underline{U}(R) := \{\underline{u}(r_i) \mid 1 \leq i \leq n\}$$

is the *collated Voronoi tessellation on S generated by R*.

Two further definitions that will be important in the following are that of the *expanded Voronoi polygon polyhedron associated with* $r_i \in R$,

$$\bar{u}(r_i) := \bigcup_{p \in r_i} \upsilon(p),$$

and that of the *boundary region of the collated polygon polyhedron associated with* r_i, which is defined to be $\bar{u}(r_i)/\underline{u}(r_i)$. The boundary regions of the collated polygons polyhedra that constitute the collated Voronoi tessellation jointly make up what is referred to as the *boundary region of the collated Voronoi tessellation*.

Before we put these definitions to use in completing Kamp and Partee's proposal discussed in the previous section, it is worth emphasizing that, thus defined, collated Voronoi tessellations have various desirable properties. For one, the cells of a collated Voronoi tessellation are provably convex, like those of ordinary Voronoi tessellations. Douven et al. (2013) also show that if all prototypical regions in a space are connected,[8] then there are no "gaps" in the boundary region of the collated Voronoi tessellation generated by them. So, for instance, it is not coincidental that there are no pockets of "white" in the black-colored boundary region of the collated Voronoi tessellation shown in Figure 8.1. That same tessellation illustrates another important property of collated Voronoi tessellations, to wit, that their boundary regions have a certain "thickness." This is important because, as Douven and coauthors note, borderline cases tend to be completely surrounded by other borderline cases (e.g., slightly changing the color of a red-orange borderline case can easily result in another red-orange borderline case; given an ordinary Voronoi tessellation, "almost all" slight adjustments would result in either a clear instance of red or a clear instance of orange).

4 From collated Voronoi tessellations to graded membership

To see how the conceptual spaces framework in the version just outlined can help fix the problem that remained for Kamp and Partee's proposal and determine a unique graded membership relation in a nonarbitrary manner, it is first to be noted how the key elements of Kamp and Partee's proposal all have natural geometric interpretations in that framework.

First, a space S with a collated Voronoi tessellation $\underline{U}(R)$ defined on it can be thought of as providing a partial model for the predicates that can be interpreted in the space (e.g., color predicates if the space is color space). Specifically, restricted collated polygons polyhedra can play the role of Kamp and Partee's extensions, the complements of expanded collated polygons polyhedra, the role of their anti-extensions, and the boundary regions contain the indeterminate instances.

Second, the ordinary Voronoi tessellations that comprise $\underline{U}(R)$ can play the role of completions. To appreciate this, note that each of those ordinary Voronoi tessellations divides the boundary region of any expanded polygon polyhedron $\bar{u}(r)$, with $r \in R$, into two, with one part containing the borderline cases that are grouped with the extension of the predicate associated with $\underline{u}(r)$ and the other part containing the remaining borderline cases. Because ordinary Voronoi tessellations are functions of similarity rankings, the way they divide the boundary region is fully in accord with the main idea underlying Kamp and Partee's distinction between those completions that respect similarity orderings (which go into their model) and those that do not.

To show how Kamp and Partee's proposal can be implemented using these geometric components, first consider again the finite case, which here means the case in which all prototypical regions in a given space consist of only finitely many points. This is the kind of case depicted in Figure 8.2, which shows a two-dimensional space S whose prototypical regions contain only two points each: $\{a, b\}, \{c, d\}, \{e, f\},$ and $\{g, h\}$. In this case, $\mathcal{V}(S)$ contains $2^4 = 16$ ordinary Voronoi tessellations. Then, supposing the geometric interpretations of the components of Kamp and Partee's proposal, and where A is a predicate designating a concept represented in this space, an item represented by point p in this space is an A to a degree that equals the number of members of $\mathcal{V}(S)$ that locate p in the polygon associated with one of the two A prototypes in the space, divided by the total number of members of $\mathcal{V}(S)$. Thus, for instance, as shown in some detail

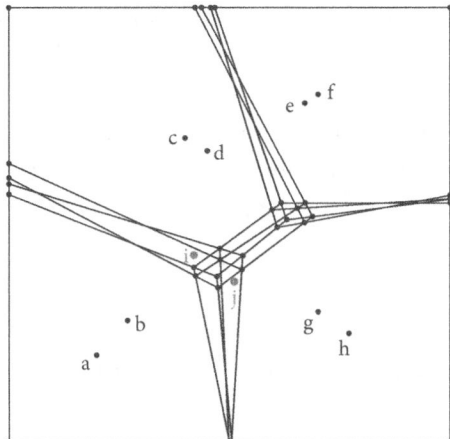

Figure 8.2 Collated Voronoi tessellation whose prototypical regions contain only finitely many points.

in Decock and Douven (2014), the point i in the space of Figure 8.2 belongs to the concept with prototypical region $\{a, b\}$ to a degree of 1/2, given that 8 of the 16 ordinary Voronoi tessellations that constitute the collated tessellation on this space place i in the polygon associated with a member of $\{a, b\}$; and similarly j belongs to the same concept to a degree of 1/4.

This generalizes straightforwardly to other conceptual spaces with finite prototypical regions. The generalization to spaces some or all of whose prototypical regions consist of *infinitely* many points is not quite so straightforward. Loosely put, the idea Decock and Douven (2014) put forward is that the above proposal can be generalized by using the fact that completions—which, as we saw, can be represented by tuples of points generating Voronoi tessellations—can themselves be represented by points in a space. For example, where we are given an m-dimensional space S with n prototypical regions, the completions can be represented by points in a different, $m \times n$-dimensional, space S^*. And, supposing we can interpret the predicate A in S, this allows us to define the degree to which a given item x is A as the Lebesgue measure of the set of points in S^* which represent completions that group the point representing x in S together with the clear A cases in S relative to the Lebesgue measure of S^* as a whole.

More formally, where S is an m-dimensional space with metric δ and prototypical regions $R = \{r_1,...,r_n\}$, we can think of each $\vec{p} = \langle p_1,...,p_n \rangle \in \Pi(R)$ as determining a completion of the partial model determined by the collated

Voronoi tessellation on S. Because for each $i \in \{1,\ldots,n\}$, we can identify p_i with the m-tuple $\langle x_{i_1},\ldots,x_{i_m} \rangle$ of its coordinates in S, completions can be identified with $m \times n$-tuples $\langle x_{1_1},\ldots,x_{1_m},\ldots,x_{n_1},\ldots,x_{n_m} \rangle$ of real numbers, which can themselves be thought of as points in an $m \times n$-dimensional space $S^* \subseteq \mathbb{R}^{m \times n}$. Then, to measure sets of completions in this space, we let I_T be the indicator function for a subset $T \subseteq S^*$, and define

$$\mu(T) := \int I_T\left(\langle x_{1_1},\ldots,x_{1_m},\ldots,x_{n_1},\ldots,x_{n_m} \rangle\right) dx_{1_1} \ldots dx_{1_m} \ldots dx_{n_1} \ldots dx_{n_m},$$

with S^* as the domain of integration. Thus, in particular,

$$\mu(\Pi(R)) = \int I_{\Pi(R)}\left(\langle x_{1_1},\ldots,x_{1_m},\ldots,x_{n_1},\ldots x_{n_m} \rangle\right) dx_{1_1} \ldots dx_{1_m} \ldots dx_{n_1} \ldots dx_{n_m}$$

$$= \int_{V_{r_1}} \ldots \int_{V_{r_n}} dx_{1_1} \ldots dx_{1_m} \ldots dx_{n_1} \ldots dx_{n_m}$$

$$= \prod_{i=1}^{n} V_{r_i}$$

is the measure of the entire set of completions, with $I_{\Pi(R)}$ being the indicator function of this set. On the reasonable assumption that, for all $i \in \{1,\ldots,n\}$, the region r_i is bounded in S, the value of $\mu(\Pi(R))$ is guaranteed to be finite.

Finally, for any item x represented by a point p in S and for any predicate A interpretable in S whose prototypical instances are collectively represented by region $r_i \in R$, the set

$$S_{p,i} := \left\{ \langle x_{1_1},\ldots,x_{1_m},\ldots,x_{n_1},\ldots,x_{n_m} \rangle \mid \delta\left(p,\langle x_{i_1},\ldots,x_{i_m} \rangle\right) < \delta\left(p,\langle x_{j_1},\ldots,x_{j_m} \rangle\right) \right.$$
$$\left. \text{for all } \langle x_{j_1},\ldots,x_{j_m} \rangle \in r_j \text{ such that } i \neq j \right\}$$

comprises all elements of $\Pi(R)$ that generate Voronoi tessellations which allocate p in the Voronoi polygon polyhedron associated with $\langle x_{i_1},\ldots,x_{j_m} \rangle \in r_i$. The measure of this set is $\mu(S_{p,i})$, and to normalize, we divide by $\mu(\Pi(R))$, which gives the degree to which x belongs to A.

Note that, given these definitions, the clear cases of A, that is, the members of the (crisp) set consisting of all x such that $\mu_A(x) = 1$, are precisely the cases represented by points lying in the restricted Voronoi polygon polyhedron associated with the prototypical A region. And what in fuzzy set theory is called the "support" of A, that is, the (crisp) set of all x for which $\mu_A(x) > 0$, are precisely the cases represented by points lying in the expanded Voronoi polygon polyhedron associated with the prototypical A region. Where two predicates are represented by different collated Voronoi polygons polyhedra in a given collated Voronoi tessellation, the intersection of their extensions consists of the x represented by a point in the intersection of the boundary regions of those polygons polyhedra, which is nonempty if and only if the polygons polyhedra are adjacent in the collated Voronoi tessellation.[9]

5 Experimental work on graded membership

It was seen how embedding Kamp and Partee's proposal in the version of the conceptual spaces framework summarized in Section 5 yields a unique membership function (in fact, one per predicate) in a nonarbitrary way. Crucially, given that framework, the core components of Kamp and Partee's proposal—partial models and completions—can be given a geometrical interpretation and can in turn, qua geometric objects, be represented in a space of the right dimensionality. The Lebesgue measure on that space is then used to define the unique graded membership function. Note that what thereby has been accomplished is an *operational* definition of graded membership: given a conceptual space, we can *calculate*, for any item x and any predicate A representable in the space, to what degree x belongs to the set of A's.

Of course, so far this is just formal modeling. The question now to be addressed is whether the model is in accordance with human judgments concerning graded membership. In the introduction, it was mentioned that fuzzy set theory was motivated by the observation that some things appear to not quite belong to a given set but also appear to not be quite outside that set. If it appears to us that x is an A to a degree of somewhere around y, is that also what follows from our model, at least in general? If not, then our model is of little value.

The general idea of how one might go about testing the model is simple enough. Take any space whose geometrical structure is known or can be determined, locate the prototypical regions in that space, and then, assuming Kamp and Partee's proposal in the conceptual spaces version, compute the degrees to which various items belong to the extensions of the predicates interpretable in the space. Next, elicit people's judgments about the degrees to which those predicates apply to the given items, and finally compare the predicted and the measured degrees.

Despite the current enthusiasm for open science, it is hard to find a conceptual space that one could use to predict, in the aforementioned way, degrees of membership. It might seem that color space is a good candidate, and indeed the structure of that space is known and available through various software packages. To be precise, there are two spaces generally considered to adequately represent human judgments of color similarity, the so-called CIELAB and CIELUV spaces, where the former is said to be preferable when the colors are perceived on paper or on cloth while the latter is said to be preferable when the colors are perceived on-screen.[10] However, the locations of the prototypical regions of colors—even of the colors corresponding to Berlin and Kay's (1969) basic color terms—in either of these spaces are largely unknown. So, at least these will first have to be determined before we can calculate, for any given shade, to what extent it is green, or blue, or red, and so on.

A first step toward charting prototypical regions in color space was taken in Douven et al. (2017). Specifically, these authors located experimentally the prototypical regions of blue and green in CIELUV space.[11] They proceeded by presenting, in a first experiment, participants with a great number of cross-sections of RGB space, asking them, in one part of the experiment, to click on any point in those cross-sections that appeared prototypically green to them, and in another part of the experiment, to click on any point that appeared prototypically blue to them, where it was randomly determined which part a participant would be shown first, and where also the cross-sections appeared in an order that was randomized per participant. Next, in a second experiment, the results from the first experiment were taken as a starting point in that they served to determine the regions in CIELUV space where the density of points that had been identified as being prototypically blue and green, respectively, was highest. From those regions many shades were sampled that were presented to participants with the question of whether these shades were typically blue or green, respectively. The convex hulls of the sets of those shades that received sufficiently many positive responses were then taken to indicate the locations of the prototypical blue and green regions in CIELUV space.

This work was actually carried out with the explicit aim of testing the conceptual spaces version of Kamp and Partee's account of graded membership. To that end, Douven and coauthors not only sought to determine where in

CIELUV space the prototypical regions of blue and green are to be found, but they also elicited from participants judgments of the degrees of membership of blue and green for each of the patches in the color series shown in Figure 8.3. Figure 8.4 shows these patches, together with the prototypical blue and green regions, in CIELUV space (in the figure, the global structure of CIELUV space is shown by thinly representing the full set of Munsell chips in the space; see Douven et al., 2017 and Jraissati and Douven, 2018 for details).

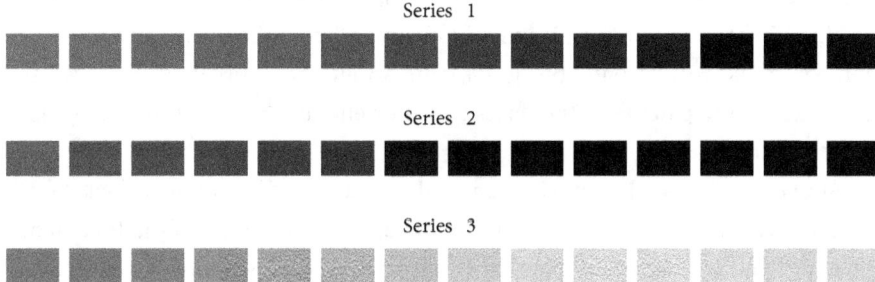

Figure 8.3 The three-color series that served as part of the materials for the experiments reported in Douven et al. (2017).

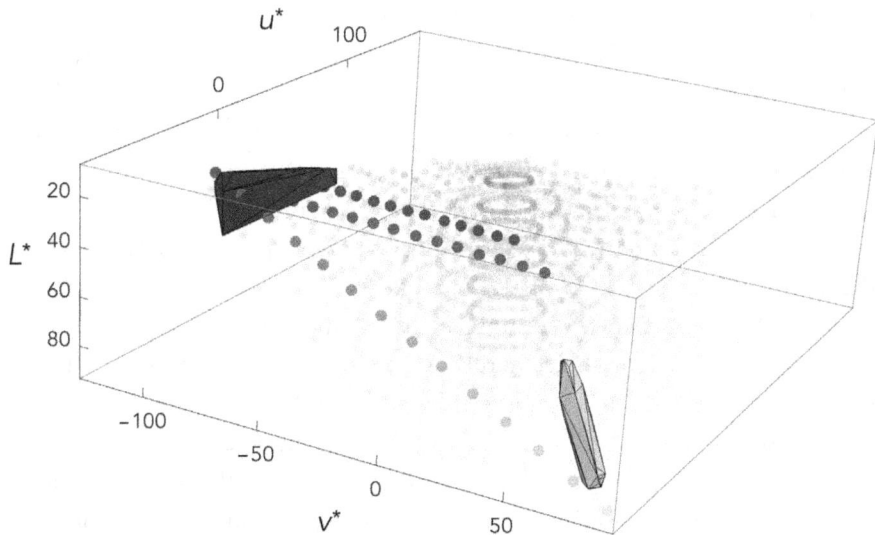

Figure 8.4 CIELUV space with Series 1–3 and prototypical regions for blue and green. Series 3 is the one with the greatest span and Series 2, which appears as the "uppermost" series, the one with the shortest span.

The color patches in Series 1–3 are all clearly in the blue-green region of color space, so we may assume that if we use the machinery from the previous section to predict their degrees of blueness and greenness on the basis of knowledge of just the prototypical blue and green regions in color space, we will obtain basically the same results we would obtain were the prototypical regions of all basic colors, or of still more colors, known. (Of course, were the prototypical regions of purple and grey known, that might make a difference for some of the patches, especially ones in Series 2, which might in some completions end up with the clear purple or the clear grey cases; see Douven et al., 2017 for more on this.) That, in any event, is what Douven and coauthors did, and they then compared the results with the judgments about the degrees of blueness and greenness of the patches in the series that they elicited from their participants in two experiments.

Above, it was said that Douven and coauthors took as locations of the prototypical blue and green regions the convex hulls of the sets of shades that had garnered enough "typically blue" and "typically green" responses, respectively. To be more exact, Douven and coauthors considered various thresholds for the number of such responses needed to qualify as a prototypical instance—the first, second, and third quartile highest-rated shades, as well as all shades that made the cut in the first experiment—and predicted degrees of blueness and greenness for the color patches in Series 1–3 on the basis of each of those thresholds separately. The observed degrees were compared with all of those different predicted degrees.

Figure 8.5 gives a graphical summary of the results of the computations as well as of the experimentally observed degrees. A number of observations can be made right away. In the top row of the figure we see that it does not matter which threshold is chosen for deriving degrees of membership from the model: all of the ones Douven and colleagues used yielded essentially the same results. Moreover, we see, in the second and third rows, that there is a close match between predicted and observed degrees of blueness, for all three series, though the match is almost perfect only for the first series. Importantly, it can be seen that the predicted degrees for any of the three series are closer to the observed degrees for that same series than they are to the observed degrees for either of the other series. Finally, the graph of the membership function is clearly S-shaped for each series, which is in line with previous empirical findings concerning graded membership (McCloskey and Glucksberg, 1979).

Douven and coauthors analyzed their data by fitting logistic curves to the observed as well as to the predicted degrees for each series, determined for each

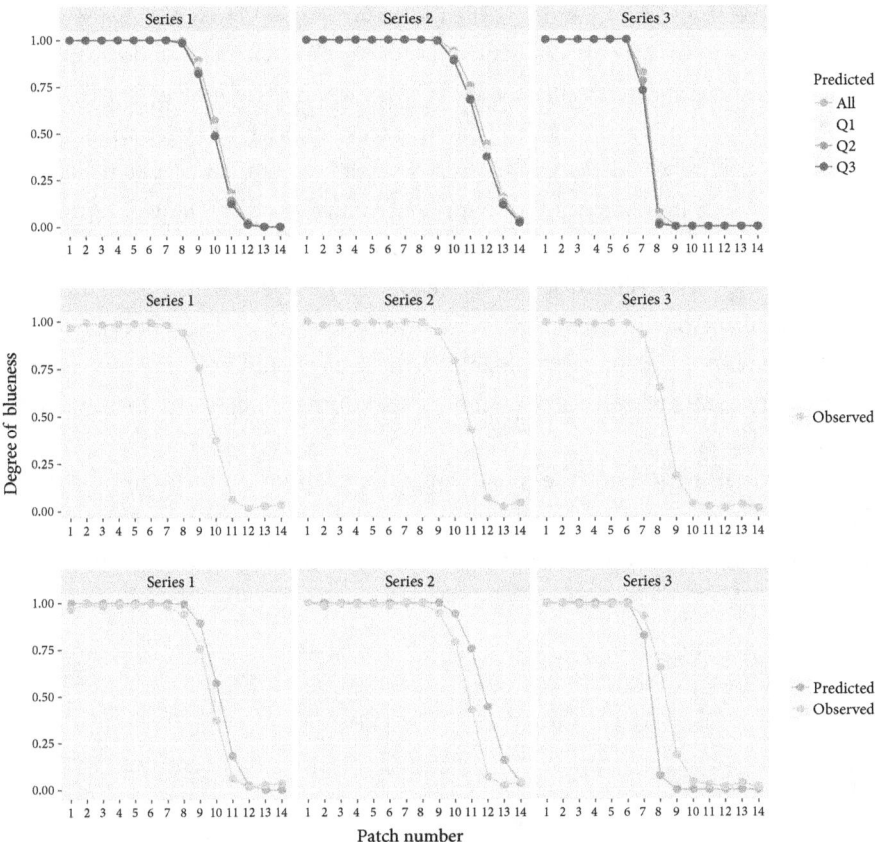

Figure 8.5 Top row showing predicted degrees of blueness for the patches in Series 1–3, assuming different thresholds for prototypicality, with Qi giving the degrees of blueness assuming as prototypical regions the convex hulls of the points with numbers of clicks greater than or equal to the i-th quartile. Middle row showing observed degrees of blueness. Bottom row showing predicted (All) versus observed degrees of blueness.

curve its so-called Point of Subjective Experience (PSE; roughly, the point at which it crosses the line $y = 0.5$), and finally calculated the slope of each curve at its PSE. The results confirmed the impression the graphs in Figure 8.5 already gave, in that the matches between predictions and observations are significantly better within series than across series. The same conclusion followed from a comparison of sums of squared deviations of predictions from observations. These experimental results were the first to buttress the conceptual spaces version of Kamp and Partee's semantics.

However, as Douven et al. (2017) were aware, the scope of their results is rather limited, given that they are confined to color space, indeed to the blue-green region in that space. Thus, those authors called for follow-up research which applies the same approach to other spaces. Douven (2016) offers one such follow-up study involving a shape space, specifically, a space for the representation of container-like objects. While Douven et al. (2017) could at least start with a space whose global structure was known, there are—as said—precious few such spaces available to researchers, and in any event the shape space considered in Douven (2016) had to be built from scratch. Thus, a number of experiments had to be carried out, for different purposes: to construct the space; once it had been constructed, to locate prototypical regions in it; and once these were known, and so relevant degrees of membership for items representable in the space could be computed, to elicit human judgments of degrees of membership, in order to compare these with the predicted degrees.

All experiments used the same forty-nine stimuli, which are shown in Figure 8.6. In a first study, over 1,000 participants were shown twenty-five pairs consisting of two different stimuli, where the pairs were randomly chosen per

Figure 8.6 The 49 figures that were used as stimuli in all experiments reported in Douven (2016). The numbers in the bottom-right square of each grid were not shown to participants; they serve here as labels to facilitate interpretation of Figure 8.7.

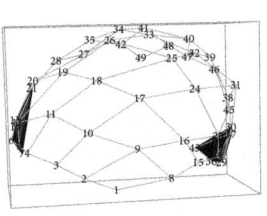 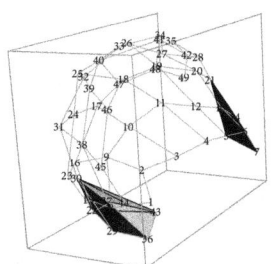 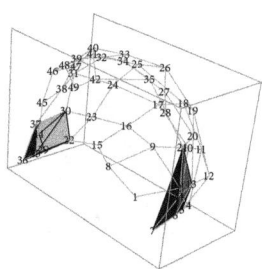

Figure 8.7 Three viewpoints of the three-dimensional city-block space, with convex hulls of majority choices of typical vase shapes (in purple) and typical bowl shapes (in green); numbers refer to the shapes in Figure 8.6.

participant. The participants were asked to rate the similarity of the members of each pair. The similarity ratings, aggregated across participants, then served as the input data for a multidimensional scaling procedure. A three-dimensional space with a city-block metric defined on it, shown in Figure 8.7, was found to do best on all relevant model-fit criteria, and also to do very well in an absolute sense; in particular, the dimensions have natural interpretations and the model has a very low stress value (see Douven, 2016 for details).

A second study sought to determine the locations of prototypical regions for concepts that could plausibly be interpreted in the space, notably the concepts of bowl, cup, mug, pot, and vase. There turned out to be very little support for the claim that any of the shapes in Figure 8.6 is typical for cups, mugs, and pots. By contrast, a majority of participants deemed various shapes typical for vases and various other shapes typical for bowls. The locations of these shapes in the three-dimensional city-block space are, together with their convex hulls, shown in purple and green, respectively, in Figure 8.7.

With the conceptual space and the prototypical regions in place, Kamp and Partee's proposal could again be used to calculate, for each of the stimuli, the degree of vasehood and degree of bowlhood. These degrees were then compared with the degrees of vasehood and bowlhood for those stimuli as judged by over 500 participants of the third study in Douven (2016). This study elicited degrees of vasehood and bowlhood in a variety of ways (the details of which need not detain us here), but it was found that however those degrees were measured, there was an exceedingly close fit, across all relevant goodness-of-fit measures, between predicted and observed degrees. Here, a visual representation may suffice, which is given in Figure 8.8.

In this figure, the leftmost points of the purple and green surfaces—where the purple surfaces are at about 1 and the green surfaces at about 0—correspond

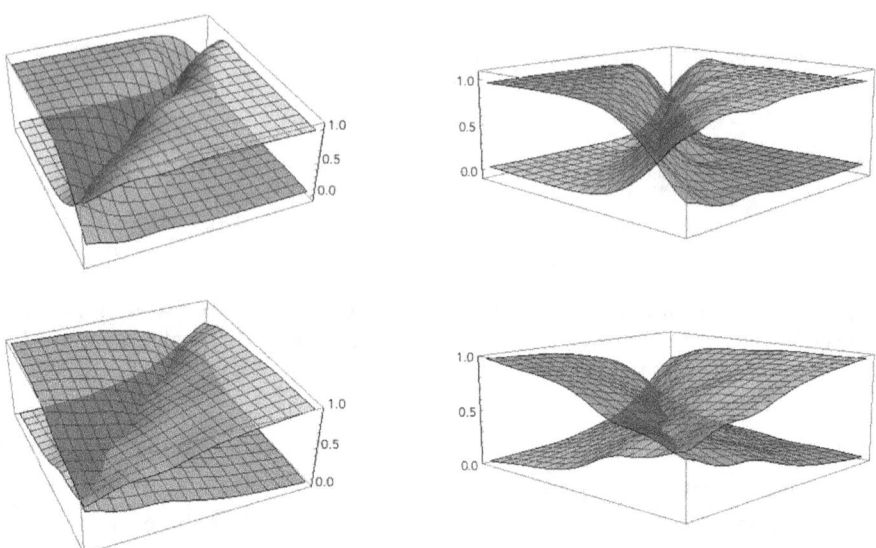

Figure 8.8 Different views on the membership functions for vase (purple) and bowl (green) for predicted (top row) and observed (bottom row) degrees of vasehood and bowlhood. See the text for explanation.

to the left upper point in Figure 8.6, specifically the shape numbered 7 in that figure; the rightmost points—where the purple surfaces are at about 0 and the green at about 1—correspond to the right lower point in Figure 8.6, specifically the shape numbered 43. Imagining the surfaces in Figure 8.8 to be projected onto Figure 8.6 in that way gives an impression of the degrees of membership accrued by the various stimuli. We see that what were identified as the prototypical vase and bowl shapes (as shown in Figure 8.7) receive the highest degrees of membership, not just according to the predictions (which is unsurprising) but also according to the observations. Furthermore, the graphs of the membership functions are again clearly S-shaped (S-shaped manifolds in this case), which as mentioned previously was expected on the basis of older research. As noted in Douven (2016), the predicted degrees over-accentuate the S-shapes of the membership functions somewhat, compared with the observed degrees, but that may be an effect of the so-called central tendency bias, which is known to pertain to tasks involving Likert scales (which were used for eliciting participants' judgments concerning the degrees of vasehood and bowlhood).[12] Most importantly, the visual impression suggests a good fit between predicted and observed degrees, in line with the results from the statistical analysis presented in Douven (2016).

6 Concluding remarks

Kamp and Partee almost succeeded in providing an operational definition of fuzzy or graded membership. The one remaining problem for their proposal—that it failed to provide a *unique* measure of membership—was solved by embedding the proposal into a special but independently motivated version of the conceptual spaces framework. Thereby, the constituent parts of Kamp and Partee's proposal could all be given geometric interpretations, and the geometry of conceptual spaces was then enough to fix a unique graded membership function (one per predicate, that is).

This was still only a first part of the response to the critique that fuzzy set theory is built on a subjective and ill-understood notion. To the extent that fuzzy set theory was fueled by the realization that, often enough, there is no clear-cut answer to the question of whether a given item belongs to a given set—often, the pre-theoretically right thing to say is that the item belongs to the set only to some degree—we still need to know whether the predictions the embedded account makes are in line with our own judgments concerning graded membership. This question was addressed in Douven (2016) and Douven et al. (2017). Together, the results reported in those papers provide successful applications of the conceptual spaces version of Kamp and Partee's proposal for modeling fuzzy membership, thereby contributing to the provision of a foundation for fuzzy set theory.

To be sure, the evidence in favor of Kamp and Partee's proposal is still limited, coming from experiments on no more than two conceptual spaces. As stated previously, there is currently no third space that could be used off the shelf to extend the results from Douven (2016) and Douven et al. (2017). There is nothing to prevent us from making additional spaces available ourselves, in the manner in which the shape space of Douven (2016) was constructed—except, perhaps, that doing so will typically require large numbers of participants (for reasons explained in Douven, 2016) and is thus likely to be rather costly.

Already with the present detailed knowledge of the two spaces discussed in the previous section—color space, and the vase-bowl shape space—we can do further empirical work to investigate the descriptive adequacy of fuzzy set theory, supposing the operationalization of its key notion summarized in Section 6. For instance, the relation of fuzzy set intersection which Zadeh formulated in his first publication on the subject, and which was stated in the introduction, was later generalized in a variety of ways, by Zadeh and others. These alternatives use so-called t-norms, three popular examples of which are the *drastic t-norm*,

$$t_D\left(\mu_{\tilde{S}_1}(\omega), \mu_{\tilde{S}_2}(\omega)\right) := \begin{cases} \min\left[\mu_{\tilde{S}_1}(\omega), \mu_{\tilde{S}_2}(\omega)\right] & \text{if } \max\left[\mu_{\tilde{S}_1}(\omega), \mu_{\tilde{S}_2}(\omega)\right] = 1 \\ 0 & \text{otherwise,} \end{cases}$$

the *Hamacher product*,

$$t_H\left(\mu_{\tilde{S}_1}(\omega), \mu_{\tilde{S}_2}(\omega)\right) := \frac{\mu_{\tilde{S}_1}(\omega) \cdot \mu_{\tilde{S}_2}(\omega)}{\mu_{\tilde{S}_1}(\omega) + \mu_{\tilde{S}_2}(\omega) - \mu_{\tilde{S}_1}(\omega) \cdot \mu_{\tilde{S}_2}(\omega)},$$

and the *Einstein product*,

$$t_E\left(\mu_{\tilde{S}_1}(\omega), \mu_{\tilde{S}_2}(\omega)\right) := \frac{\mu_{\tilde{S}_1}(\omega) \cdot \mu_{\tilde{S}_2}(\omega)}{2 - \left(\mu_{\tilde{S}_1}(\omega) + \mu_{\tilde{S}_2}(\omega) - \mu_{\tilde{S}_1}(\omega) \cdot \mu_{\tilde{S}_2}(\omega)\right)}.$$

While so far such proposals have been defended on *a priori* grounds, the combined use of color space and the vase-bowl shape space allows us to compare these proposals empirically and to tell which, if any, accords most closely with how humans aggregate fuzzy sets.

For example, consider the shape numbered 18 in Figure 8.6 and imagine this to have the color of the seventh patch in Series 3 shown in Figure 8.3. The shape was predicted and observed to be a borderline vase-bowl case, and the color was predicted and deemed to be borderline blue-green. Now to what degree is the imagined shape a blue vase? Assuming Zadeh's original definition of fuzzy set intersection, we predict that it is the minimum of the degree to which the shape is a vase and the degree to which it is blue (given that the latter is around 0.75, whichever of the results shown in Figure 8.5 we look at, while consulting the data from Douven, 2016 shows that the former is between 0.49 and 0.58, the prediction would be between 0.49 and 0.58). On the other hand, assuming the drastic t-norm, the answer would be 0; after all, neither the degree of vasehood nor the degree of blueness is 1 in this case. And assuming the Hamacher or the Einstein product, the answer would be somewhere between 0.4 and 0.5 or somewhere between 0.3 and 0.4, respectively. Here, the experiment to carry out is straightforward enough: show participants the shapes in Figure 8.6, but now colored using the colors from Series 1–3 in Figure 8.3; ask them to rate the degree to which they deem the figures blue vases (or green bowls, etc.); and finally compare the responses with the predictions that can be derived from the results in Douven (2016) and Douven et al. (2017) in conjunction with the above and possibly various other proposals for defining fuzzy set intersection.

There is no guarantee that the results will support any of the extant fuzzy set intersection operations. Nor is there a guarantee that any such operation will cover all possible combinations of fuzzy sets, or will cover them equally well in all contexts. But this just goes to show that the project of providing foundations for fuzzy set theory is truly empirical in nature, and as such has only just begun.[13]

Notes

1. All figures in this chapter are printed in grayscale. Color versions of those figures for which color is relevant can be found at https://osf.io/vkb5e/.
2. Water and washing powder are still needed though, even with the new logic.
3. So-called epistemicists (e.g., Williamson, 1994) would disagree with the characterization of borderline cases I am giving here. For my misgivings about epistemicism, see Keefe (2000: Ch. 3) and Burgess (2001) for compelling arguments against epistemicism.
4. What Kamp and Partee call "completions" are in the literature on vagueness more commonly referred to as "precisifications"; given that Kamp and Partee's account of vagueness will be central to the foundations for fuzzy set theory I am proposing in this paper, I will stick to their terminology throughout.
5. See the first four chapters of Gärdenfors (2000) for an excellent introduction to the conceptual spaces framework.
6. Whether they would have led to roughly the *same* shape space is an open question that can only be settled by conducting the said experiment.
7. A region is convex if and only if, for any two points lying in the region, the whole line segment connecting the points lies in the region as well. See Sivik and Taft (1994), Regier et al. (2007), and Jraissati and Douven (2017, 2018) for empirical findings suggesting that our concepts correspond to convex regions in similarity spaces. See Douven and Gärdenfors (2019) and Douven (2019b) for why we should expect our concepts to be represented by convex regions.
8. A region is connected if for any two points in the region there is a path connecting them which lies itself entirely in the region. Note that any convex region is necessarily connected, but not the other way round, given that connectedness does not require the paths to be line segments.
9. That there is such a clean split between cases that fully belong to a set—those that receive a degree of membership of 1—and those that do not might raise concerns

about the possibility of second-order vagueness. See Douven and Decock (2017) on how to address such concerns in the present framework. The same paper also shows how to define degrees of truth, and indeed to obtain a full-blown semantics for a language with vague predicates, on the basis of degrees of membership.
10 See Malacara (2002) or Fairchild (2013). But see also Douven (2017) for some reasons to believe that CIELUV space is generally preferable.
11 Given that all their experiments were carried out online, CIELUV space seemed the appropriate space to work with.
12 See Stevens (1971) and Douven (2018) for more on the central tendency bias.
13 I am greatly indebted to Christopher von Bülow and to two anonymous referees for valuable comments on a previous version.

References

Berlin, B. and Kay, P. (1969). *Basic Color Terms*. Stanford, CA: CSLI Publications.

Burgess, J. (2001). Vagueness, epistemicism, and response-dependence. *Australasian Journal of Philosophy*, 79:507–24.

Decock, L. and Douven, I. (2014). What is graded membership? *Noûs*, 48: 653–82.

Douven, I. (2016). Vagueness, graded membership, and conceptual spaces. *Cognition*, 151:80–95.

Douven, I. (2017). Clustering colors. *Cognitive Systems Research*, 45:70–81.

Douven, I. (2018). A Bayesian perspective on Likert scales and central tendency. *Psychonomic Bulletin and Review*, no. 25:1203–11.

Douven, I. (2019a). Are all types of implicatures equally natural? A multidimensional scaling perspective. Manuscript.

Douven, I. (2019b). The rationality of vagueness. In R. Dietz (Ed.), *Vagueness and Rationality*. New York: Springer, in press.

Douven, I. and Decock, L. (2017). What verities may be. *Mind*, 126:386–428.

Douven, I., Decock, L., Dietz, R. and Égré, P. (2013). Vagueness: A conceptual spaces approach. *Journal of Philosophical Logic*, 42: 137–60.

Douven, I. and Gärdenfors, P. (2019). What are natural concepts? A design perspective. Mind & Language, forthcoming.

Douven, I., Wenmackers, S., Jraissati, Y. and Decock, L. (2017). Measuring graded membership: The case of color. *Cognitive Science*, 41:686–722.

Dubois, D. and Prade, H. (1979). Fuzzy real algebra: Some results. *Fuzzy Sets and Systems*, 2:327–48.

Dubois, D. and Prade, H. (1982). Towards fuzzy differential calculus. *Fuzzy Sets and Systems*, 8:1–17; 105–16; 225–33.

Fairchild, M. D. (2013). *Color Appearance Models*. Hoboken, NJ: Wiley.

Fine, K. (1975). Vagueness, truth and logic. *Synthese*, 30:265–300.

Gärdenfors, P. (2000). *Conceptual Spaces*. Cambridge, MA: MIT Press.
Gärdenfors, P. (2014). *The Geometry of Meaning*. Cambridge, MA: MIT Press.
Halmos, P. (1974). *Measure Theory*. New York: Springer.
Jraissati, Y. and Douven, I. (2017). Does optimal partitioning of color space account for universal color categorization? *PLoS One*, 12:e0178083. https://doi.org/10.1371/journal.pone.0178083.
Jraissati, Y. and Douven, I. (2018). Delving deeper into color space. Manuscript.
Kamp, H. and Partee, B. (1995). Prototype theory and compositionality. *Cognition*, 57:129–91.
Keefe, R. (2000). *Theories of Vagueness*. Cambridge: Cambridge University Press.
Kölbel, M. (2010). Vagueness as semantic. In R. Dietz and S. Moruzzi (Eds.), *Cuts and Clouds*, 304–26. Oxford: Oxford University Press.
Lindley, D. V. (2004). Comment. *Journal of the American Statistical Association*, 99:877–79.
Malacara, D. (2002). *Color Vision and Colorimetry: Theory and Applications*, Bellingham, WA: SPIE Press.
McCloskey, M. E. and Glucksberg, S. (1979). Natural categories: Well defined or fuzzy sets? *Memory and Cognition*, 6:462–72.
Osherson, D. N. and Smith, E. E. (1981). On the adequacy of prototype theory as a theory of concepts. *Cognition*, 29:259–88.
Regier, T., Kay, P. and Khetarpal, N. (2007). Color naming reflects optimal partitions of color space. *Proceedings of the National Academy of Sciences USA*, 104:1436–41.
Rosch, E. (1973). Natural categories. *Cognitive Psychology*, 4:328–50.
Sivik, L. and Taft, C. (1994). Color naming: A mapping in the IMCS of common color terms. *Scandinavian Journal of Psychology*, 35:144–64.
Stevens, S. S. (1971). Issues in psychophysical measurement. *Psychological Review*, 78:426–50.
Unger, P. (1980). The problem of the many. *Midwest Studies in Philosophy*, 5:411–67.
Williamson, T. (1994). *Vagueness*. London: Routledge.
Zadeh, L. (1965). Fuzzy sets. *Information and Control*, 8:338–53.

9

What Isn't Obvious about "Obvious": A Data-Driven Approach to Philosophy of Logic

Moti Mizrahi

1 Introduction

What kinds of insights can we gain by analyzing a corpus of logical and philosophical texts? In this chapter, I explore one way in which the methods of data science can be used to analyze a large corpus of logical and philosophical texts in order to shed light on logical, philosophical, and metaphilosophical questions. I illustrate this data-driven approach with an example from logic and the philosophy of logic. In particular, I set out to test the way in which the oft-repeated thesis that "logic is obvious" is reflected in logical and philosophical practice. I hypothesize that, if logicians and philosophers of logic subscribe to the view that logic is obvious, then we should expect to find them using "obvious" in contexts in which they signal the certainty of the claims and/or arguments they make. If they use "obvious" when their claims and/or arguments betray a lack of certainty, then that would suggest that their use of "obvious" should not be taken literally and that their view, as evinced by what they do (more precisely, the arguments they put forward), is that logic is not obvious.[1]

To put it another way, I think that the text mining, analysis, and visualization techniques of data science can help us address the following question: What would logic, philosophical logic, and philosophy of logic look like *in practice* if practitioners subscribed to the view that logic is obvious? Instead of picking out a few examples of logicians and philosophers of logic who have asserted that logic itself, or some proper subset of logic, is obvious, and deriving a general lesson from such examples, I examine a large corpus of text and look for correlations that are designed to pick out such assertions on a much larger scale. My hope is that this data-driven methodology can be used to investigate systematically other claims that are often made about fields like logic and philosophy as well as other disciplines.²

For the purposes of this empirical study, I have focused on the idea that logic is obvious. Many logicians and philosophers of logic have expressed approval of the thesis that logic is obvious. In a recent post on the blog of the American Philosophical Association, Steven M. Chan (2017) relates the following story about Willard van Orman Quine:

> Once in [an introduction to symbolic logic] course, after he wrote a proof on the board, a student raised his hand and asked impatiently, "Why bother writing out that proof? It's obvious." To which Quine replied, "Young man, this entire course is obvious."

Quine expresses this idea explicitly in print when he says that "every logical truth is obvious" (Quine, 1970: 82) and that "elementary logic is obvious or can be resolved into obvious steps" (Quine, 1976: 112). Other logicians and philosophers of logic share this view that "logic is a theory of the obvious" (Sher, 1999: 207). According to Murawski (2014: 87), the "axioms and rules of logic are true in an obvious way," and Shenefelt and White (2013: 301) say that "a good deal of logic is, in fact, so intuitively obvious that it needs no further justification." In addition to being obvious, logic is also said to be "necessary and a priori" (Russell, 2015: 793).³ Even those who do not subscribe to the idea that logic is obvious, necessary, and a priori claim that, both in principle and in practice, "good students in an intro logic class may regard classical logic, not just as a reasonable theory, but as *obviously correct*" (Russell, 2015: 797; emphasis in original). The point, then, is that the word "obvious" gets used quite frequently when logicians and philosophers of logic describe logic or engage in the philosophy of logic. Moreover, standard logic texts make liberal use of "obvious" in describing proofs, theorems, consequences, axioms, rules, etc. For instance, "Double negation is fairly obvious and needs little explanation" (Hurley, 2012: 402).

As Sher (2010: 158) points out, however, "the idea that logic is obvious is a vague idea." The word "obvious" itself is vague in more ways than one. According to the *Oxford English Dictionary*, "obvious" is used as either an adjective or a noun in the following ways:

1. Plain and evident to the mind; perfectly clear or manifest; plainly distinguishable; clearly visible.
2. Lacking in subtlety, sophistication, or originality; banal, predictable.
3. Natural, likely; such as common sense might suggest.

In addition to these different ways in which "obvious" can be used, various things can be said to be obvious. For example:

1. Propositions can be said to be obvious in the sense of being *evident to the mind*. Explanations can be said to be obvious in the sense of being *perfectly clear*. Shades of color can be said to be obvious in the sense of being *plainly distinguishable*. Macroscopic objects, like trees and cars, can be said to be obvious in the sense of being *clearly visible* to a sensory apparatus.
2. Punchlines can be said to be obvious in the sense of *lacking in subtlety or sophistication*. Theses can be said to be obvious in the sense of *lacking in originality*. Consequences can be said to be obvious in the sense of being *predictable*.
3. Events can be said to be obvious in the sense of being *natural* or *likely*. Decisions can be said to be obvious in the sense of being *what common sense might suggest*.

Accordingly, the meaning of "obvious" will depend on the ways in which the word "obvious" is used and the things that are said to be obvious. In other words, *context* partly determines the meaning of "obvious." In this chapter, then, I set out to examine the contexts in which the word "obvious" is used in logic, philosophical logic, and the philosophy of logic. Through a systematic examination of these contexts, I aim to test empirically how the idea that "logic is obvious" is reflected in logical and philosophical practice. My approach is data-driven. That is to say, I propose that systematically searching for patterns of usage in databases of scholarly works, such as JSTOR, can provide new insights into the ways in which the idea that logic is obvious is reflected in logical and philosophical practice, that is, in the arguments that logicians and philosophers of logic actually make in their scholarly publications.[4]

2 Methods

Most logic textbooks instruct students to look for indicator words when trying to identify arguments in texts. For example, according to Salmon (2013: 39), indicator words are

> Words commonly used to signal premisses or conclusions of arguments. Examples of premiss indicator words are *for*, *since*, *because*, and *for the reason that*. Examples of conclusion indicator words are *hence*, *thus*, *therefore*, *and so*, *it follows that*, and *for that reason*. (Emphasis in original)[5]

Indicator words are also supposed to help in distinguishing between deductive and inductive arguments. For example, according to Baronett (2016: 23):

> To help identify arguments as either *deductive* or *inductive*, one thing we can do is look for key words or phrases. For example, the words "necessarily," "certainly," "definitely," and "absolutely" suggest a deductive argument. . . . On the other hand, the words "probably," "likely," "unlikely," "improbable," "plausible," and "implausible" suggest inductive arguments. (emphasis added)

Likewise, according to Hurley (2012: 34), "inductive indicators" include terms and phrases such as "probably," "improbable," "plausible," "implausible," "likely," "unlikely," and "reasonable to conclude," whereas "deductive indicators" include terms and phrases such as "it necessarily follows that," "certainly," "absolutely," and "definitely."

Accordingly, I propose to use indicator words to test how the idea that "logic is a theory of the obvious" (Sher, 1999: 207) and that "logic is [...] necessary and a priori" (Russell, 2015: 793) is reflected in the scholarly work of logicians and philosophers of logic. If logicians and philosophers of logic take logic to be obvious, necessary, and a priori, then the arguments they actually make in practice, that is, in scholarly publications, should reflect that in some way. More explicitly, my empirical study is designed to address the following questions:

- Does "obvious" correlate with deductive indicators, such as "necessarily" and "certainly"?
- Does "obvious" correlate with inductive indicators, such as "probably" and "likely"?
- Does "obvious" correlate with hedging markers, such as "suggest" and "seem"?

I propose that, if logicians and philosophers of logic accept the idea that logic is obvious, necessary, and a priori, then this idea would be reflected in actual practice, that is, in the way in which logicians and philosophers of logic *do* and *talk* about logic in argumentative writings, as follows:

A. If practitioners understand logic as obvious, necessary, and a priori, then we would expect to find positive correlations between the word "obvious" and deductive indicators like "necessarily" and "certainly."
B. If practitioners understand logic as obvious, necessary, and a priori, then we would expect to find negative correlations (or no correlations at all) between the word "obvious" and inductive indicators like "probably" and "likely."

C. If practitioners understand logic as obvious, necessary, and a priori, then we would expect to find negative correlations (or no correlations at all) between the word "obvious" and hedging markers like "suggest" and "seem."

In other words, if logicians and philosophers of logic subscribe to the view that logic is obvious, necessary, and a priori, then we would expect to see them make the sort of arguments that reflect that when they do logic or talk about logic. That is to say, we would see instances of "obvious" occur mostly in the context of deductive arguments, rather than inductive arguments, since deductive arguments are said to be "indefeasible" and are the sort of arguments whose premises "provide conclusive support for their conclusions" or "necessitate the truth of the conclusion" (Sinnott-Armstrong and Fogelin, 2015: 181).[6]

To be clear, I am not suggesting that logicians and philosophers of logic make only deductive arguments. They probably make all kinds of arguments in their scholarly work. What I am suggesting, however, is that, if one believes that logic is obvious, necessary, and a priori, then that belief should be reflected in the sort of arguments one makes when one does logic or talks about logic. Now, since deductive arguments are the sort of arguments whose premises "provide conclusive support for their conclusions" or "necessitate the truth of the conclusion" (Sinnott-Armstrong and Fogelin, 2015: 181), we would expect practitioners who think that logic is obvious, necessary, and a priori, to make such arguments more often than other kinds of arguments (e.g., inductive arguments) when they do logic or talk about logic. In other words, if practitioners subscribe to the view that logic is obvious, necessary, and a priori, then we would expect to see more deductive arguments than other kinds of argument in logical and philosophical practice in the context of talk about the obviousness of logic.[7]

The data driving this empirical study is taken from JSTOR Data for Research (jstor.org/dfr). This database allows researchers to search full texts for exact phrases and access the metadata associated with search results. I used this database to search for the term "obvious," as well as "obviously" and "obviousness" (JSTOR allows for truncation or "wildcard" search using "obvious*"), through research articles written in English. JSTOR does not have a discipline category for logic in particular, so I have created a dataset from the five logic journals in the JSTOR database, namely, *The Journal of Symbolic Logic* (1936–2012), *Studia Logica* (1953–2012), *Journal of Philosophical Logic* (1972–2012), *Journal of Logic, Language, and Information* (1992–2012), and *The Bulletin of Symbolic Logic* (1995–2012).

Table 9.1 Deductive, inductive, and hedging indicator pairs

Deductive indicator pairs	Inductive indicator pairs	Hedging indicator pairs
Therefore necessarily	Therefore probably	Therefore seem
Therefore certainly	Therefore likely	Therefore suggest
Follows necessarily	Follows probably	Follows seem
Follows certainly	Follows likely	Follows suggest

Having selected the journals for data mining, I then searched for "obvious*" in the context of deductive indicators, inductive indicators, and hedging markers through research articles contained in my Logic dataset, and ran statistical analyses on search results relative to their proportions in the JSTOR corpus in order to test empirically how the idea that logic is obvious is manifested in logical and philosophical practice. To make sure that the aforementioned indicator words occur in the contexts of arguments, to rule out nonargumentative (e.g., rhetorical) instances of "obvious" as much as possible, and to test predictions (A)–(C) above, I have anchored them to the argument indicators "therefore" and "follows" within ten words of each other (using the operator ~10 in JSTOR's search box). This method yields the deductive indicator pairs and inductive indicator pairs listed in Table 9.1.

Here is an example of "obvious" and the argument indicator "therefore" that this search methodology would pick out (emphasis added):

> The possibility of situations with the classical valuation of conjunction, disjunction and conditional is *obvious. Therefore* . . . (Kovač, 2012: 347)

Likewise, here is an example of "obvious" and the argument indicator "follows" that this search methodology would pick out (emphasis added):

> *Obviously*, every admissible numbering x is computable. Moreover, we have that $\beta \leq x$. Since Z is dense in Q_c, *it follows that* Q_c is recursively separable (Spreen, 1998: 195).

I have also anchored the hedging markers "seem" and "suggest" to the argument indicators "therefore" and "follows" to make sure that my search methodology will pick out usage of these hedging markers in argumentative contexts.[8] This method yields the following hedging indicator pairs: "therefore seem," "follows seem," "therefore suggest," and "follows suggest" (see Table 9.1).

This search methodology will pick out instances of these indicator words such as the following (emphasis added):

> *Obviously*, Γ and Π are **Z**-specific. . . . either the antecedent is not **Z**-specific, or the succedent is not *necessarily* irreducible (Petersen, 2000: 396).

Unlike modal logic . . . and thus the elimination of variables is not *likely* to proceed. . . . Models can be built over the set C in the *obvious* way (Kracht, 2013: 1338).

Partial logic (in which some statements receive no truth value) *suggests* itself as an *obvious* candidate . . . (Horsten, 2015: 687).

For the purposes of comparison, I ran the same searches I did on the Logic dataset on three discipline categories in the JSTOR database: philosophy, mathematics, and biological sciences. I have selected the philosophy discipline category on JSTOR because it is useful to compare it to the Logic dataset. Most of the journals contained in JSTOR's philosophy category publish work in so-called mainstream areas of philosophy, by professionally trained philosophers, that purport to offer arguments, which should, in theory, contain many indicator words. The so-called "general" philosophy journals, such as *Mind* and *Noûs*, often publish work on logic, philosophical logic, and the philosophy of logic. In fact, some of the most influential papers in logic and the philosophy of logic, such as Tarski (1944), Turing (1950), Quine (1951), and Davidson (1967), were published in so-called "general" philosophy journals, such as *Philosophy and Phenomenological Research*, *Mind*, *Philosophical Review*, and *Synthese*, which are included in JSTOR's philosophy category.

Much like philosophy, mathematics is supposed to be like logic in terms of being a priori (cf. Ashton and Mizrahi, 2018b), which is why I have selected JSTOR's mathematics discipline category for the sake of comparison as well. I have also selected the biological sciences category in JSTOR because biology is supposed to be different from logic insofar as the latter is supposed to be a priori, whereas the former is supposed to be a posteriori.

3 Results

Before testing predictions (A)–(C), it would be useful to get a sense of how widespread "obvious" talk is in research articles on logic, philosophical logic, and the philosophy of logic, and to compare that to how often "obvious" is used in our comparison disciplines, namely, philosophy, mathematics, and biology. So I first looked at the number of Logic, Philosophy, Mathematics, and Biology research articles that contain the word "obvious" and its cognates ("obviously" and "obviousness") in the JSTOR corpus.[9] As mentioned above, JSTOR allows for truncation or "wildcard" search, so I used the search term "obvious*."

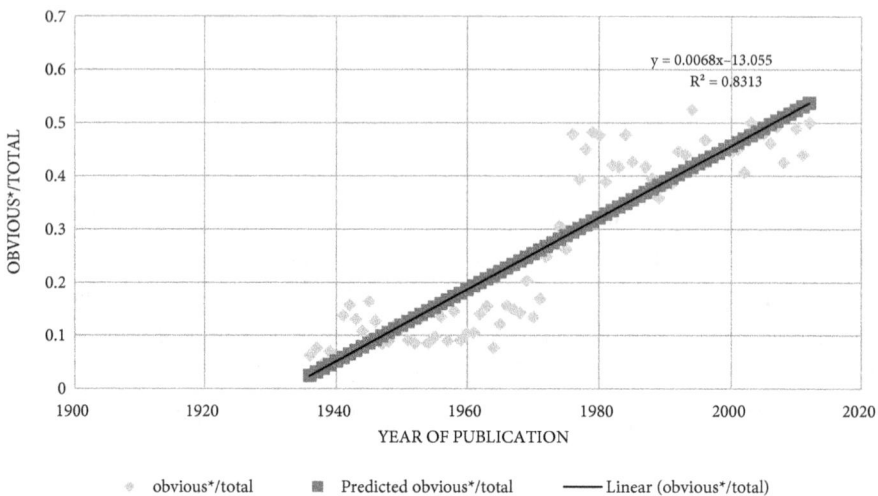

Figure 9.1 Proportions of research articles in the Logic dataset that contain "obvious" and its cognates.
Source: JSTOR Data for Research.

Figure 9.1 shows the proportions of research articles (written in English) in the Logic dataset that contain the word "obvious" and its cognates (from 1936 until 2012).

As far as the Logic dataset is concerned, a linear regression analysis indicates that year predicts proportion of "obvious" talk in logic research articles. A significant regression equation was found ($F(1, 75) = 369.58, p < .001$), with an R^2 of .83. This suggests that "obvious" talk in logic research articles is becoming significantly more widespread over the years.

Figure 9.2 shows the proportions of research articles (written in English) in the Philosophy dataset that contain the word "obvious" and its cognates (from 1900 until 2012).

As far as the Philosophy dataset is concerned, a linear regression analysis indicates that year predicts proportion of "obvious" talk in philosophy research articles. A significant regression equation was found ($F(1, 111) = 271.57, p < .001$), with an R^2 of .71. This suggests that "obvious" talk in philosophy research articles is becoming significantly more widespread over the years.

Figure 9.3 shows the proportions of research articles (written in English) in the Mathematics dataset that contain the word "obvious" and its cognates (from 1900 until 2012).

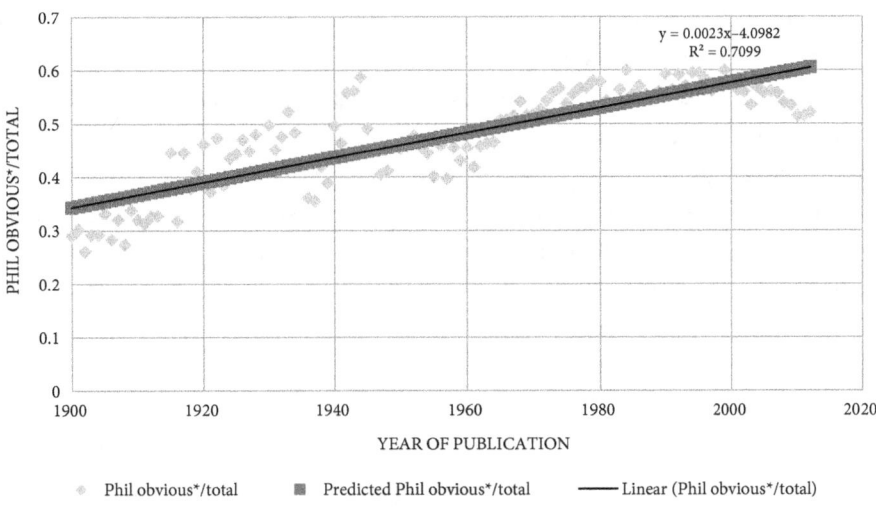

Figure 9.2 Proportions of research articles in the Philosophy dataset that contain "obvious" and its cognates.
Source: JSTOR Data for Research.

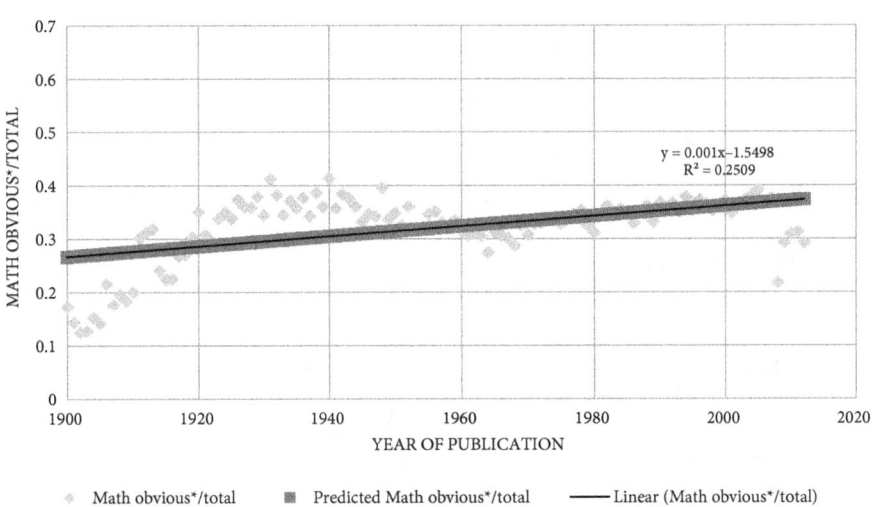

Figure 9.3 Proportions of research articles in the Mathematics dataset that contain "obvious" and its cognates.
Source: JSTOR Data for Research.

As far as the Mathematics dataset is concerned, a linear regression analysis indicates that year predicts proportion of "obvious" talk in mathematics research articles. A significant regression equation was found ($F(1, 111) = 37.19, p < .001$), with an R^2 of .25. This suggests that "obvious" talk in mathematics research articles is becoming significantly more widespread over the years.

Finally, Figure 9.4 shows the proportions of research articles (written in English) in the Biology dataset that contain the word "obvious" and its cognates (from 1900 until 2012).

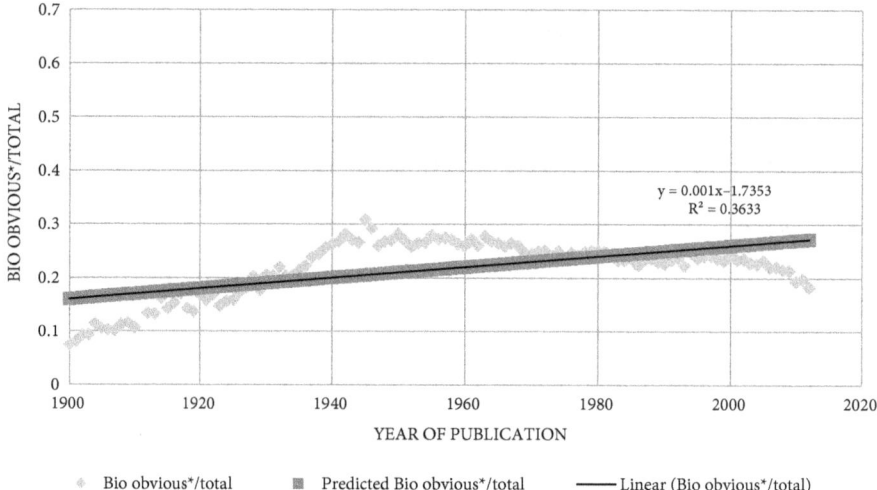

Figure 9.4 Proportions of research articles in the Biology dataset that contain "obvious" and its cognates.

Source: JSTOR Data for Research.

Table 9.2 Mean proportions of research articles that contain the word "obvious" and its cognates relative to the total number of research articles in the Logic (1936–2012), Philosophy (1900–2012), Mathematics (1900–2012), and Biology (1900–2012) datasets

	Mean	*SD*	*N*
Philosophy	.48	.09	112
Mathematics	.33	.06	112
Logic	.29	.16	77
Biology	.22	.05	112

As far as the Biology dataset is concerned, a linear regression analysis indicates that year predicts proportion of "obvious" talk in biology research articles. A significant regression equation was found ($F(1, 111) = 63.35, p < .001$), with an R^2 of .36. This suggests that "obvious" talk in biology research articles is becoming significantly more widespread over the years.

Overall, these results suggest that "obvious" talk is quite widespread in logical and philosophical practice (Table 9.2) and that it is becoming significantly more widespread over the years. As we have seen, in Philosophy and in Logic, year explains 71 percent and 83 percent of the variation in proportion of "obvious" talk in research articles, respectively, whereas in Mathematics and Biology it explains 25 percent and 36 percent, respectively.

Now that we have a sense of how widespread "obvious" talk is in Logic, Philosophy, Mathematics, and Biology, we can turn to testing predictions (A)–(C). If practitioners indeed understand logic as obvious, necessary, and a priori, we would also expect instances of "obvious" to occur in the context of deductive arguments, since unlike inductive arguments, deductive arguments are the sort of arguments that are said to be "indefeasible" and whose premises "provide conclusive support for their conclusions" or "necessitate the truth of the conclusion" (Sinnott-Armstrong and Fogelin, 2015: 181). For example, if one thinks, as Mittelstaedt and Stachow (1978: 184) do, that, in the calculus Q_{eff} of effective quantum logic, "it is *obvious* that, without further knowledge about the mutual commensurability of *a* and *b*, it is impossible to decide whether the proposition $a \wedge b$ is true or false" (emphasis added), then it would be rather odd if one were to preface the conclusion that "it is impossible to decide whether the proposition $a \wedge b$ is true or false" with "probably," "likely," or "it seems," for this conclusion follows necessarily from what is taken to be obvious here. In general, then, if logicians and philosophers of logic subscribe to the view that logic is obvious, necessary, and a priori, then we would expect to see them make deductive arguments more often than other kinds of argument when they do logic or talk about logic, for deductive arguments are the sort of arguments whose conclusions follow necessarily from their premises.

To find out if this is indeed the case, I tested for correlations between "obvious" and the deductive indicator pairs, inductive indicator pairs, and hedging indicator pairs listed in Table 9.1 in the Logic, Philosophy, Mathematics, and Biology datasets. The results are summarized in Table 9.3.

Table 9.3 Pearson correlation coefficients for "obvious" and the deductive, inductive, and hedging indicator pairs in the Logic, Philosophy, Mathematics, and Biology datasets

	Indicator Pairs	Logic	Philosophy	Mathematics	Biology
Deductive	Obvious* & therefore necessarily	.47	.18	.01	.25
	Obvious* & therefore certainly	.49	−.21	.18	.05
	Obvious* & follows necessarily	.48	.21	.46	−.23
	Obvious* & follows certainly	.42	.04	.03	−.24
Inductive	Obvious* & therefore probably	−.01	−.03	−.10	.63
	Obvious* & therefore likely	.31	.52	.37	.31
	Obvious* & follows probably	.15	.01	−.03	.01
	Obvious* & follows likely	.17	.51	.24	.49
Hedging	Obvious* & therefore seem	.37	−.26	−.05	.01
	Obvious* & therefore suggest	.29	.55	.36	.32
	Obvious* & follows seem	.43	.37	.20	−.33
	Obvious* & follows suggest	.39	.55	.16	.40

3.1 "Obvious*" and deductive indicator pairs

As mentioned in Section 2, if practitioners understand logic as obvious, necessary, and a priori, we would expect "obvious" to positively correlate with deductive indicator pairs, such as "therefore necessarily" and "follows certainly." In other words, if practitioners understand logic as obvious, necessary, and a priori, then we would expect them to make arguments that reflect that obviousness in the form of deductive arguments whose conclusions *follow necessarily* from their premises.

As we can see from Table 9.3, in the Logic dataset, "obvious*" is positively correlated with the deductive indicator pairs "therefore necessarily" ($r = .47$), "therefore certainly" ($r = .49$), "follows necessarily" ($r = .48$), and "follows certainly" ($r = .42$), as we would expect if logicians and philosophers of logic subscribe to the idea that logic is obvious, necessary, and a priori. In the Philosophy dataset, however, things are rather more complicated in practice than they are in theory. For the deductive indicator pair "therefore certainly" is negatively correlated with "obvious*" ($r = -.21$). Given that the Pearson correlation coefficient r can tell us about the linear relationship between two variables (positive or negative) and the strength of that relationship (the closer r is to 0, the weaker the linear relationship; the closer r is to -1, the stronger the negative relationship; the closer r is to 1, the stronger the positive relationship), we can say that the correlations between "obvious*" and the deductive indicator pairs "therefore necessarily" ($r = .18$), "follows necessarily" ($r = .21$), and "follows certainly" ($r = .04$) are rather weak as far as the Philosophy dataset is concerned. Indeed, the positive correlation between the deductive indicator pair "therefore necessarily" and "obvious*" is stronger in Biology ($r = .25$) than in Philosophy ($r = .18$), which is contrary to what we would expect if unlike logic and philosophy, which are supposed to be a priori, biology is supposed to be a posteriori. Although mathematics is supposed to be like logic in terms of being a priori, and "obvious*" is positively correlated with deductive indicator pairs in the Mathematics dataset, these positive correlations are generally weaker than those in the Logic dataset, with the exception of "follows necessarily" ($r = .46$).

If practitioners understand logic as obvious, necessary, and a priori, we would expect the correlations between "obvious*" and deductive indicator pairs to be stronger in Philosophy and Logic than in Biology. Although the deductive indicator pairs "follows necessarily" and "follows certainly" are negatively correlated with "obvious*" in Biology, the deductive indicator pair "therefore certainly" is positively correlated with "obvious*" in Biology, but not in Philosophy, and the correlation between the deductive indicator pair "therefore necessarily" and "obvious*" in Philosophy is weaker than that in Biology. This result is not quite what we would expect to find in logical and philosophical practice because, if practitioners subscribe to the idea that logic is obvious, necessary, and a priori, then logical and philosophical practice should reflect that with language closer to that of mathematical practice than biological practice. For, unlike logic, philosophy, and mathematics, which are supposed to be a priori, biology is supposed to be a posteriori. That is, deductive indicator pairs should be positively correlated with "obvious*" at least as strongly as "follows necessarily" is correlated with "obvious*" in Mathematics ($r = .46$). In the Logic dataset, "obvious*" is positively correlated with all the deductive indicator pairs and somewhat more strongly than in Mathematics.

In terms of prediction (A), then, the results do meet our expectations overall. Prediction (A) is borne out by the data to the extent that there are positive correlations between "obvious*" and the deductive indicator pairs in Logic. However, there is also a negative correlation between "obvious*" and a deductive indicator pair in Philosophy, and the positive correlations are rather weak, sometimes even weaker than those in a supposedly a posteriori discipline like biology in which conclusions are tentative, not obvious.

3.2 "Obvious*" and inductive indicator pairs

As mentioned in Section 2, if practitioners understand logic as obvious, necessary, and a priori, we would expect "obvious" to not only positively correlate with deductive indicator pairs, such as "therefore necessarily" and "follows certainly," but also to negatively correlate (or not correlate at all) with inductive indicator pairs, such as "therefore probably" and "follows likely." In other words, if practitioners understand logic as obvious, necessary, and a priori, then we would expect them to make arguments that reflect that, namely, deductive arguments whose premises necessarily entail their conclusions, not inductive arguments that provide probable, but not conclusive, support for their conclusions.

As we can see from Table 9.3, as far as research articles in logic, philosophical logic, and the philosophy of logic are concerned, "obvious" talk is positively correlated with inductive indicator pairs, with the exception of "therefore probably" ($r = -.01$ in Logic and $r = -.03$ in Philosophy), though these negative correlations are rather weak. In Mathematics, the inductive indicator pairs "therefore probably" and "follows probably" are negatively correlated with "obvious*" ($r = -.10$ and $r = -.03$), as we might expect, since mathematics, like logic and philosophy, is supposed to be a priori, but "obvious*" is positively correlated with "therefore likely" ($r = .37$) and "follows likely" ($r = .24$). In Philosophy, however, the inductive indicator pairs, with the exception of "therefore probably," are positively correlated with "obvious*." The strength of the positive correlations between "obvious*" and the inductive indicator pairs "therefore likely" ($r = .52$) and "follows likely" ($r = .51$) approximates that between "obvious*" and inductive indicator pairs in Biology ($r = .63$ and $r = .49$), as does the strength of the positive correlation between "obvious*" and "therefore likely" in Logic ($r = .31$). In that respect, logical and philosophical practice look more like biological practice than mathematical practice, which is contrary to what we would expect if logicians and philosophers of logic take logic to be obvious, necessary, and a priori, given that logic, like mathematics and philosophy, is supposed to be a priori. For, if they take logic to be obvious, necessary, and a priori, then we would expect to see logicians and philosophers of logic make the sort of arguments whose premises *necessitate* (rather than make more probable) the truth of their conclusions. In other words, we would expect to see logicians and philosophers of logic make *deductive*, rather than inductive, arguments.

If practitioners understand logic as obvious, necessary, and a priori, we would expect the correlations between "obvious*" and inductive indicator pairs to be stronger in Biology than in Philosophy or Logic, since biology is supposed to be a posteriori, whereas logic and philosophy are supposed to be a priori. As we have seen, however, that is not quite what we find in practice. With the exception of "therefore probably," which is negatively correlated with "obvious*" in Philosophy ($r = -.03$), the other inductive indicator pairs are not only positively correlated with "obvious*" in Philosophy but the strength of these correlations approximates that of positive correlations in Biology. Moreover, "obvious" talk in Logic is also positively correlated with the inductive indicator pairs with strength that approximates that of a positive correlation in Biology. Contrary to expectations, these results suggest that, as far as scholarly work in logic, philosophical logic, and the philosophy of logic is concerned, "obvious" talk is likely to occur not only in the context of deductive argumentation but also in the context of inductive argumentation. These results are at odds with the way in which we would expect to see the idea that logic is obvious reflected in logical and philosophical practice because the conclusions of inductive arguments are probable, not certain, and thus always tentative or doubtful to some extent, whereas obvious and necessary truths are supposed to be neither tentative nor doubtful.

In terms of prediction (B), then, the results are rather mixed. Prediction (B) is partially borne out by the data to the extent that there is a negative correlation between "obvious*" and an inductive indicator pair in both Logic and Philosophy. However, there are more positive than negative correlations between "obvious*" and inductive indicator pairs in both Logic and Philosophy, and those positive correlations are often as strong as those in a supposedly a posteriori discipline like biology in which conclusions are tentative, not obvious or necessary, whereas a supposedly a priori discipline like mathematics shows some negative correlations between "obvious*" and inductive indicator pairs.

3.3 "Obvious*" and hedging indicator pairs

As mentioned in Section 2, if practitioners understand logic as obvious, necessary, and a priori, we would expect "obvious" to not only positively correlate with deductive indicator pairs, such as "therefore necessarily" and "follows certainly," and negatively correlate (or not correlate at all) with inductive indicator pairs, such as "therefore probably" and "follows likely," but also to negatively correlate (or not correlate at all) with hedging indicator pairs, such as "therefore seem" and "follows suggest." After all, there is no need to hedge an obvious, necessary, and a priori truth. For instance, if a statement like "All squares are rectangles" is obvious, necessary, and a priori, then there is no need to preface it with "it seems that," as in "It seems that all squares are rectangles." Likewise, if "double negation is fairly obvious and needs little explanation" (Hurley, 2012: 402), then it would be rather odd to preface it with "evidence suggests that," as in "Our evidence suggests that $p \equiv \sim \sim p$."

As we can see in Table 9.3, all the hedging indicator pairs are positively correlated with "obvious*" in the Logic dataset. These results suggest that logical practice is rather like biological practice, given that the hedging indicator pairs "therefore suggest" ($r = .29$ in Logic and $r = .32$ in Biology) and "follows suggest" ($r = .39$ in Logic and $r = .40$ in Biology) are positively correlated with "obvious*" in Biology, as one might expect, since biology, unlike logic, is supposed to be an a posteriori discipline in which conclusions are tentative, not obvious or necessary. In Philosophy, too, the hedging indicator pairs, with the exception of "therefore seem" ($r = -.26$), are positively correlated with "obvious*." Indeed, these correlations between hedging indicator pairs and "obvious*" are generally stronger in Philosophy than in Biology. Contrary to expectations, then, these results show that hedging indicator pairs, such as "therefore suggest" ($r = .29$ in Logic and $r = .55$ in Philosophy) and "follows seem" ($r = .43$ in Logic and $r = .37$ in Philosophy), are positively correlated with "obvious*," thus suggesting that logicians and philosophers of logic hedge their claims in argumentative contexts when they do logic or talk about the obviousness of logic.

If logicians and philosophers of logic take logic to be obvious, necessary, and a priori, we would expect the correlations between "obvious*" and hedging indicator pairs to be stronger in Biology than in Logic or in Philosophy, given that biology is supposed to be a posteriori, where conclusions are tentative, whereas logic and philosophy are supposed to be a priori, where conclusions are obvious and necessary. As we have seen, however, that is not quite what we find in practice. With the exception of "therefore seem," which is negatively correlated with "obvious*" in Philosophy ($r = -.26$), the other hedging indicator pairs are not only positively correlated with "obvious*" in Philosophy but also more strongly correlated than in Biology. In fact, in the Logic dataset, we see positive correlations between "obvious*" and all the hedging indicator pairs that are generally stronger than positive correlations in Biology. These results are at odds with the way in which we would expect to see the idea that logic is obvious reflected in logical and philosophical practice because there is no need to hedge claims that are obvious, necessary, and a priori. Again, if a statement like "All squares are rectangles" is obvious, necessary, and a priori, then it would be odd to say "It seems that all squares are rectangles" or "Our evidence suggests that all squares are rectangles." Even though, on the idea that "logic is a theory of the obvious" (Sher, 1999: 207), logical truths are supposed to be like "All squares are rectangles" insofar as they are supposed to be obvious, necessary, and a priori, the results from the Logic dataset suggest that logicians and philosophers of logic use hedging markers when they make such statements in argumentative contexts.

In terms of prediction (C), then, the results are somewhat mixed as well. While there is a negative correlation between "obvious*" and a hedging indicator pair in Philosophy, which is supposed to be a priori like logic, most of the correlations between "obvious*" and hedging indicator pairs are positive in both Logic and Philosophy. In fact, the Logic dataset shows positive correlations between "obvious*" and the hedging indicator pairs that are generally stronger than positive correlations in Biology, even though biology is supposed to be different from both logic and philosophy in terms of being an a posteriori field with tentative conclusions rather than an a priori field with obvious and necessary conclusions.

4 Discussion

The results of my empirical study can be summed up as follows. As we would expect to see reflected in logical practice (i.e., in research articles on logic, philosophical logic, and philosophy of logic) if practitioners subscribe to the view that "logic is a theory of the obvious" (Sher, 1999: 207) and that "logic is ... necessary and a priori" (Russell, 2015: 793):

1. The word "obvious" (and its cognates) is positively correlated with the deductive indicator pairs "therefore necessarily" ($r = .47$ in Logic and

$r = .18$ in Philosophy), "therefore certainly" ($r = .49$ in Logic), "follows necessarily" ($r = .48$ in Logic and $r = .21$ in Philosophy), and "follows certainly" ($r = .42$ in Logic and $r = .04$ in Philosophy).
2. The word "obvious" (and its cognates) is negatively correlated with the inductive indicator pair "therefore probably" in Logic ($r = -.01$) and in Philosophy ($r = -.03$).
3. The word "obvious" (and its cognates) is negatively correlated with the hedging indicator pair "therefore seem" in Philosophy ($r = -.26$).
4. Generally, the strongest positive correlations between "obvious" talk and the deductive indicator pairs are found in the Logic dataset.

Contrary to what we would expect to see reflected in logical practice (i.e., in research articles on logic, philosophical logic, and philosophy of logic) if practitioners subscribe to the view that "logic is a theory of the obvious" (Sher, 1999: 207) and that "logic is . . . necessary and a priori" (Russell, 2015: 793):

5. Although positive, the correlations between the word "obvious" (and its cognates) and deductive indicator pairs in Philosophy are rather weak.
6. The word "obvious" (and its cognates) is negatively correlated with the deductive indicator pair "therefore certainly" in Philosophy ($r = -.21$).
7. The word "obvious" (and its cognates) is positively correlated with the inductive indicator pairs "therefore likely" ($r = .31$ in Logic and $r = .52$ in Philosophy), "follows probably" ($r = .15$ in Logic and $r = .01$ in Philosophy), and "follows likely" ($r = .17$ in Logic and $r = .51$ in Philosophy).
8. The word "obvious" (and its cognates) is positively correlated with the hedging indicator pairs "therefore seem" ($r = .37$ in Logic), "therefore suggest" ($r = .29$ in Logic and $r = .55$ in Philosophy), "follows seem" ($r = .43$ in Logic and $r = .37$ in Philosophy), and "follows suggest" ($r = .39$ in Logic and $r = .55$ in Philosophy).
9. Generally, the positive correlations between "obvious" talk and the inductive indicator pairs are as strong in Philosophy as they are in Biology.
10. Generally, the positive correlations between the word "obvious" (and its cognates) and the hedging indicator pairs are as strong in Logic as they are in Biology.

These mixed results do not warrant any definitive conclusions about the status of the idea that logic is obvious in logical practice. Indeed, these mixed results suggest that the idea that logic is obvious is not reflected in any straightforward way in the arguments that logicians and philosophers of logic actually make in their scholarly work. Instead, they suggest that things are rather more complicated in practice than they are in theory. For although many logicians and philosophers of logic subscribe to the idea "logic is a theory of the obvious" (Sher, 1999: 207) and that "logic is . . . necessary and a priori" (Russell, 2015: 793), as evidenced by the fact that "obvious" talk is becoming significantly more widespread in logic (Figure 9.1) and philosophy (Figure 9.2) over the years, the arguments they actually put forward in their scholarly work appear in the context of inductive indicators and hedging markers. This is contrary to the sort of arguments we would expect to see in research articles on logic, philosophical logic, and philosophy of logic if practitioners took logic to be obvious, necessary, and a priori.

For this reason, I submit, the results of my empirical study raise an interesting question that is worthy of further investigation: if logicians and philosophers of logic subscribe to the view that logic is obvious, necessary, and a priori, why do they hedge their claims and frame their arguments using inductive indicators? The fact that they do so in practice is prima facie in tension with the idea that logic is obvious, necessary, and a priori. This fact, then, calls for an explanation. Providing such an explanation is beyond the scope of this chapter, so I hope that future studies will be able to shed light on this issue.[10]

Beyond the aforementioned results, I think that my empirical study speaks to the broader question of how studying philosophical texts from a data science perspective can shed new light on logical, philosophical, and metaphilosophical questions. While some of the data I have mined from the JSTOR database do indeed meet our expectations, there are some surprising findings as well. One surprising finding, as mentioned above, is that "obvious" talk is becoming significantly more widespread in logic (Figure 9.1) and philosophy (Figure 9.2) over the years, and that such talk also occurs in the context of inductive argumentation and hedging markers. This suggests that, although they subscribe to the view that logic is obvious, logicians and philosophers of logic engage in inductive argumentation and hedge their claims when they do logic or talk about the obviousness of logic. This finding could not have come to light without the data mining and corpus analysis techniques of data science. This, then, is an illustration of how data science can help us gain new insights about logic, philosophy, and other disciplines as well, when we apply its methods to analyzing large corpora of scholarly work in those fields of study.

My hope is that this data-driven methodology can be used to investigate systematically other claims that are often made about fields like logic and philosophy as well as other disciplines.[11] As more refined techniques of data mining and analysis become available, such that they would allow us to conduct more focused studies using specialized corpora, we will be able to gain a better understanding of the ways in which practitioners in philosophy, mathematics, logic, and biology conceive of their arguments and methods. In that respect, the application of data science to logic, philosophical logic, and the philosophy of logic can help initiate a discussion on the methodology of logic and philosophy of logic just as the application of social science to philosophy, also known as "experimental philosophy," has engendered a fruitful discussion about philosophical methodology in recent years. As Dolcini (2017: 102) puts it, "The advent of experimental philosophy . . . revitalized the discussion about a major meta-philosophical issue: what are the proper methods, aims and ambitions of philosophy?" Data science, I propose, can do for logic, philosophical logic, and the philosophy of logic in particular what experimental philosophy did for philosophy in general.

5 Conclusion

In this chapter, I have looked at one way in which a data science approach to studying logical and philosophical texts can shed new light on logical, philosophical, and metaphilosophical questions. In particular, I have tested empirically how the idea that logic is obvious, necessary, and a priori is reflected in logical and philosophical practice. The results of my empirical survey of data mined systematically from the JSTOR database suggest that there is a difference between theory and practice as far as the idea that logic is obvious is concerned. That is, although my results suggest that logicians and philosophers of logic subscribe to the view that logic is obvious, necessary, and a priori *in theory*, given that "obvious" talk is becoming significantly more widespread in logic and philosophy over the years, they also show that, *in practice*, "obvious" talk in logic and philosophy often occurs in the context of inductive argumentation and hedging markers. This is contrary to the sort of arguments we would expect to see in research articles on logic, philosophical logic, and the philosophy of logic if practitioners understand logic to be obvious, necessary, and a priori.

These findings, I submit, raise the following interesting question for further research: if logicians and philosophers of logic subscribe to the view that logic is obvious, necessary, and a priori, as evidenced by the fact that "obvious" talk is becoming significantly more widespread in both logic and philosophy over the years, why do they hedge their claims and frame their arguments using inductive indicators when they use the term "obvious" and its cognates? I have proposed that addressing such questions using the methods of data science can help initiate a discussion on methodology in logic and the philosophy of logic in much the same way that addressing philosophical questions using the methods of social science (i.e., "experimental philosophy") has led to a fruitful discussion about philosophical methodology.

Acknowledgments

I would like to thank Andrew Aberdein and Matthew Inglis for inviting me to contribute to this volume and to two anonymous reviewers for helpful comments on an earlier draft of this chapter.

Notes

1 Attempts to incorporate empirical or experimental methods into philosophy have often been met with the "that's not philosophy" (Jenkins, 2014) or "that's not philosophically significant" charge. (See, for example, Kauppinen, 2007. Cf. Knobe, 2007; O'Neill and Machery, 2014.) This issue is beyond the scope of this chapter. For present purposes, I hope it is enough to point out that the volume for which this chapter was invited is about empirical or experimental approaches to logic and/or philosophy of logic and mathematics. In that respect, this chapter offers an empirical approach to a question about logical and philosophical practice.
2 See, for example, Ashton and Mizrahi (2018b) for an empirical, data-driven investigation of how the idea that philosophy is a priori is reflected in philosophical practice.
3 According to Field (1998: 4), "It makes sense to regard logic as a priori but at the same time to think it conceivable that further conceptual developments could show it not to be, by showing that logic is empirically defeasible (in interesting ways) after all." Cf. Bueno (2010: 105) who argues that fallibilism and apriorism are not reconcilable.
4 As an anonymous reviewer pointed out, what one considers obvious might sometimes be left unsaid. As we will see, "obvious" talk is quite widespread in logic

and philosophy journal articles and is becoming significantly more widespread over the years.

5   See also Bessie and Glennan (2000: 4), Copi et al. (2011: 11–12), and Marcus (2018: 10).
6   According to Augustus De Morgan (1839: 3), "The question of logic is, does the conclusion certainly follow if the premises be true?"
7   See, for example, Suppes's (1999 :129) reason for not including mathematical content in the practice exercises for chapter 4 of his logic textbook: "It is often intuitively *obvious that the conclusions logically follow from the premises* given in the exercises" (emphasis added).
8   On the hedging uses of "seem" in philosophy, see Cappelen (2012: 46). On both "seem" and "suggest" as hedging markers in corpus linguistics and natural language processing, see Thabet (2018: 679–81).
9   When "logic" is not capitalized, it refers to the field of study or academic discipline of logic. When "Logic" is capitalized, it refers to the dataset of logic articles mined from the JSTOR database. The same applies to "philosophy" (field) and "Philosophy" (dataset), "mathematics" (field) and "Mathematics" (dataset), and "biology" (field) and "Biology" (dataset).
10 In that respect, it is interesting to point out the increase in "obvious" talk around the 1970s (see Figure 9.1) and the fact that the *Journal of Philosophical Logic* (JPL) published its first issue in 1972. Further studies are needed to determine whether, and the extent to which, JPL is responsible for the significant increase in "obvious" talk in logic as a whole. Thanks to Andrew Aberdein for this point.
11 For another example of applying the methods of data science to philosophy, see Ashton and Mizrahi (2018a).

References

Ashton, Z. and Mizrahi, M. (2018a). Intuition talk is not methodologically cheap: Testing the "received wisdom" about armchair philosophy. *Erkenntnis*, 83(3): 595–612.

Ashton, Z. and Mizrahi, M. (2018b). Show me that argument: Empirically testing the armchair philosophy picture. *Metaphilosophy*, 49 (1–2):58–70.

Baronett, S. (2016). *Logic*. New York, NY: Oxford University Press.

Bessie, J. and Glennan, S. (2000). *Elements of Deductive Inference*. Belmont, CA: Wadsworth.

Bueno, O. (2010). Is logic a priori? *The Harvard Review of Philosophy*, 17(1):105–17.

Cahn, S. M. (2017). How teachers succeed. *Blog of the APA*, May 11. Retrieved from http://blog.apaonline.org/2017/05/11/how-teachers-succeed/

Cappelen, H. (2012). *Philosophy without Intuitions.* Oxford: Oxford University Press.
Copi, I. M., Cohen, C. and McMahon, K. (2011). *Introduction to Logic,* fourteenth edition. New York, NY: Prentice Hall.
Davidson, D. (1967). Truth and meaning. *Synthese,* 17(3):304–23.
De Morgan, A. (1839). *First Notions of Logic (Preparatory to the Study of Geometry).* London: Printed for Taylor and Walton, Booksellers and Publishers to University College.
Dolcini, N. (2017). Philosophy made visual: An experimental study. In L. Magnani and Casadio, C. (Eds.). *Model-Based Reasoning in Science and Technology: Logical, Epistemological, and Cognitive Issues,* 101–18. Basel: Springer.
Field, H. (1998). Epistemological nonfactualism and the a prioricity of logic. *Philosophical Studies,* 92(1):1–24.
Horsten, L. (2015). One hundred years of semantic paradox. *Journal of Philosophical Logic,* 44(6):681–95.
Hurley, P. J. (2012). *A Concise Introduction to Logic,* eleventh edition. Boston, MA: Wadsworth.
Jenkins, K. (2014). "That's not philosophy": Feminism, academia, and the double bind. *Journal of Gender Studies,* 23(3):262–74.
Kauppinen, A. (2007). The rise and fall of experimental philosophy. *Philosophical Explorations,* 10(2):95–118.
Knobe, J. (2007). Experimental philosophy and philosophical significance. *Philosophical Explorations,* 10(2):119–21.
Kovač, S. (2012). Logical opposition and collective decisions. In J. Béziau and D. Jacquette (Eds.), *Around and Beyond the Square of Opposition,* 341–56. Basel: Springer.
Kracht, M. (2013). Are logical languages compositional? *Studia Logica* 101(6):1319–40.
Marcus, R. (2018). *Introduction to Formal Logic with Philosophical Applications.* New York, NY: Oxford University Press.
Mittelstaedt, P. and Stachow, E. W. (1978). The principle of excluded middle in quantum logic. *Journal of Philosophical Logic,* 7(1):181–208.
Murawski, R. (2014). *The Philosophy of Mathematics and Logic in the 1920s and 1930s in Poland,* Translated from Polish by Maria Kantor. Basel: Springer.
O'Neill, E. and Machery, E. (2014). Experimental philosophy: What is it good for? In E. Machery and E. O'Neill (Eds.), *Current Controversies in Experimental Philosophy,* vii–xxix. New York, NY: Routledge.
Petersen, U. (2000). Logic without contraction as based on inclusion and unrestricted abstraction. *Studia Logica,* 64(3):365–403.
Quine, W. V. (1951). Two dogmas of empiricism. *Philosophical Review,* 60(1):20–43.
Quine, W. V. (1970). *Philosophy of Logic.* Englewood Cliffs, NJ: Prentice Hall.
Quine, W. V. (1976). *The Ways of Paradox and Other Essays,* Revised and enlarged edition. Cambridge, MA: Harvard University Press.

Russell, G. (2015). The justification of the basic laws of logic. *Journal of Philosophical Logic*, 44(6):793–803.

Salmon, M. H. (2013). *Introduction to Logic and Critical Thinking*, sixth edition. Boston, MA: Wadsworth.

Shenefelt, M. and White, H. (2013). *If A, then B: How the World Discovered Logic*. New York, NY: Columbia University Press.

Sher, G. (1999). Is logic a theory of the obvious? *European Review of Philosophy*, 4:207–38.

Sher, G. (2010). Epistemic friction: Reflections on knowledge, truth, and logic. *Erkenntnis*, 72(2):151–76.

Sinnott-Armstrong, W. and Fogelin, R. J. (2015). *Understanding Arguments*, Ninth edition. Stamford, CT: Cengage Learning.

Spreen, D. (1998). On effective topological spaces. *The Journal of Symbolic Logic*, 63(1):185–221.

Suppes, P. (1999). *Introduction to Logic*. Mineola, NY: Dover Publications, Inc.

Tarski, A. (1944). The semantic conception of truth: And the foundations of semantics. *Philosophy and Phenomenological Research*, 4(3):341–76.

Thabet, R. A. (2018). A cross-cultural corpus study of the use of hedging markers and dogmatism in postgraduate writing of native and non-native speakers of English. In K. Shaalan, A. E. Hassanien and F. Tolba (Eds.), *Intelligent Natural Language Processing: Trends and Applications*, 677–710. Gewerbestrasse: Springer.

Turing, A. M. (1950). Computing machinery and intelligence. *Mind*, 59(236):433–60.

10

Philosophy and the Psychology of Conditional Reasoning

David Over and Nicole Cruz

Conditionals and conditional reasoning are fundamental in all studies of reasoning, in logic, philosophy, and psychology. Any inference, from *A* (perhaps a long conjunction of premises) to a conclusion *B* can be summed up in the conditional, *if A then B*, and any conditional *if A then B* can be presented as an argument from *A* to *B*. Philosophers have long recognized the importance of conditionals and have studied them since classical times (Kneale and Kneale, 1962: 128–38). Psychological investigations of conditionals have a much more recent origin, and we will argue that the psychology of reasoning made a false start by taking bivalent truth functional logic as the standard of correct reasoning with the natural language indicative conditional. The most influential early psychological account of conditionals, mental model theory (Johnson-Laird and Byrne, 1991), proposed that these conditionals have the same models as the truth functional conditional. We cover philosophical objections to this position, and psychological evidence against it, and introduce the new Bayesian / probabilistic approach to the study of conditionals in the psychology of reasoning (Elqayam and Over, 2013; Oaksford and Chater, 2007; Over and Cruz, 2018). This approach has its philosophical origins in the works of Ramsey (1926/1990, 1929/1990) and de Finetti (1936/1995, 1937/1964), and their analysis of subjective probability judgments as expressions of degrees of belief. It brings with it a new logic, called the logic of probability by de Finetti and of partial belief by Ramsey, unifying the studies of probability judgment and reasoning. We will focus mainly on indicative conditionals, but will end with some remarks about counterfactuals.

1 The mental model theory of conditionals

There are many experimental results in psychology that should be of some interest to logicians and philosophers working on conditionals (Evans and Over, 2004; Over and Cruz, 2018), but the theoretical understanding of conditionals in psychology has lagged far behind that in logic and philosophy. Until relatively recently, the theory of mental models, as proposed by Johnson-Laird and Byrne (1991), was the most influential psychological account of the natural language indicative conditional, *if A then B*. According to this theory, people reason by building mental representations of the logical possibilities that make a statement true. For example, a person's full representation of an inclusive disjunction, *A or B*, would be represented by three models that made the disjunction true: one in which both *A* and *B* are true, one in which *A* is false and *B* is true, and one in which *A* is true and *B* is false. In mental model theory, a full representation of a natural language conditional, *if A then B*, was identical with the three models for the disjunction, *not-A or B*.

Johnson-Laird and Byrne (1991: 7) introduced their position using the following conditional:

(1) If Arthur is in Edinburgh (*E*), then Carol is in Glasgow (*G*).

They stated that (1) is true supposing *E* and *G* are true, Arthur is in Edinburgh, and Carol is in Glasgow, and false supposing *E* is true and *G* is false, Arthur is in Edinburgh, and Carol is not in Glasgow. They then asked, "But, suppose its antecedent is false, i.e. Arthur is *not* in Edinburgh, is the conditional true or false?" Their answer was, "It can hardly be false, and so since the propositional calculus allows only truth or falsity, it must be true." Here Johnson-Laird and Byrne assumed, with no argument, that truth functional logic, the propositional calculus, gives the correct logic and meaning of the natural language indicative conditional, with the result that this conditional is supposed to be logically equivalent to the truth functional *material conditional* of that logic, $E \supset G$, which is logically equivalent to *not-E or G*. The truth table for the material conditional is given in Table 10.1.

Table 10.1 The truth table for the material conditional $E \supset G$, equivalent to *not-E or G*

G E	1	0
1	1	0
0	1	1

1 = true and 0 = false.

Johnson-Laird and Byrne did not fully take into account the arguments of logicians and philosophers who had long pointed out serious problems with claiming that natural language conditionals are equivalent to material conditionals. Stalnaker (1968) and Lewis (1973) were prominent among these. The problems are usually summed up in the phrase "the paradoxes of the material conditional." It is logically valid to infer the material conditional *A ⊃ B*, equivalent to *not-A or B*, from *not-A*, and from *B*, but "paradoxes" follow from holding that natural language conditionals, *if A then B*, can be inferred from *not-A*, and from *B*. For example, suppose we have reason to believe it highly probable that Arthur is not in Edinburgh: he has broken his legs in an accident, and he cannot go anywhere unless Carol also goes along to help him. Assuming (1) is a material conditional, we must infer that (1) is highly probable when *not-E or G* is highly probable. But it is clearly unjustified to infer confidently in this example that, if Arthur is in Edinburgh, then Carol is in Glasgow, merely because we are highly confident that Arthur, with his broken legs, is not in Edinburgh. This is an example in which we would rather judge that (1) is improbable, implying that (1) cannot be logically equivalent to a material conditional.

Johnson-Laird and Byrne (1991: 74) argued that inferring *if A then B* from *not-A* is logically valid, but that people dislike this inference because it "throws semantic information away" (see also Orenes and Johnson-Laird, 2012). The conclusion of the inference rules out fewer logical possibilities, and so is less informative, than its premise. The authors claimed on the same basis that people also dislike the logically valid inference from *A* to *A or B*, technically called *or-introduction*. We agree that there are pragmatic reasons why speakers in a discussion would not openly assert only *A or B* when they knew *A*: making this assertion can seriously mislead their hearers (Grice, 1989). However, we have argued that pragmatic considerations will have less effect when people make inferences from their own subjective degrees of belief, and have confirmed a prediction that people will endorse inferring *A or B* from *A* in their subjective probability judgments, but not inferring *if A then B* from *not-A* (Cruz et al., 2015; Cruz et al., 2017; Politzer and Baratgin, 2016; see also Gilio and Over, 2012). This finding is also at odds with a recent radical revision of mental model theory, in which both inferences, inferring *if A then B* from *not-A*, and inferring *A or B* from *A*, are held to be invalid (Johnson-Laird et al., 2015; see Baratgin et al., 2015, and Over and Cruz, 2018, for further points about the radical revision of mental model theory).

2 The Ramsey test and conditional probability

Stalnaker (1968) and Lewis (1973) proposed non-truth functional, modal logical systems for the natural language conditional. Their accounts have had a great impact in logic and philosophy (Edgington, 1995, 2014). Simplifying these modal analyses, we can say that a conditional *if A then B* is taken to be true in the actual state of affairs if and only if B is true in the possible world (or worlds) in which A is true that is closest to the actual world.

Consider in more detail how Stalnaker's theory applies to (1) in the example we have given. His theory draws on a suggestion made by Ramsey (1929/1990) for evaluating a conditional, *if A then B*, which has become known as the *Ramsey test*. As extended by Stalnaker, one conducts the "test" by hypothetically supposing A, while making minimal changes to preserve consistency, and then assessing one's degree of confidence in B under the supposition. Making minimal changes is considering the closest possibility, the closest *possible world* in more technical terms, in which A holds, in order to see whether B holds there as well. Applying this procedure to (1) in our example, in which we believe that Arthur has broken legs and is probably not in Edinburgh, we hypothetically suppose that he has gone to Edinburgh, but still has broken legs. We then infer with high confidence that Carol has gone along to help him and so is not in Glasgow, and hence that (1) is false, and not true, as a material conditional would be. Table 10.2 represents the true (1) and false (0) evaluations of a Stalnaker conditional in the example just given, when we take ourselves to be in the *not-E & not-G* cell, while judging that the *E & not-G* cell is the closest possibility in which E holds.

Stalnaker (1970) argued that, on his account, the probability of a natural language conditional, *P(if A then B)*, is the conditional probability of B given A, $P(B \mid A)$. The Ramsey test, as we have seen, ends in a degree of confidence in B given A, and that is the conditional probability of B given A. The philosophical position that $P(if\ A\ then\ B) = P(B \mid A)$ has been simply called *the Equation*

Table 10.2 The table for the Stalnaker conditional in the *if E then G* example

G E	1	0
1	1	0
0	0	0

1 = true and 0 = false.

(Edgington, 1995; Oaksford and Chater, 2007; Over and Cruz, 2018). The name is justified because this proposed identity has such deep consequences for the study of natural language conditionals. It implies that these conditionals are not truth functional, and that the psychological studies of conditional reasoning and of probability judgment and decision making can be unified and integrated in the new Bayesian / probabilistic approaches to the study of conditionals and human cognition more generally (Elqayam and Over, 2013; Oakford and Chater, 2007, 2013). A conditional that satisfies the Equation has been termed a *probability conditional* (Adams, 1998), and a *conditional event* (de Finetti, 1936/1995, 1937/1964).

Lewis (1976) proved, however, that the Equation fails for conditionals in systems like his and Stalnaker's, which are not, then, probability conditionals or conditional events. The basic point can be illustrated by introducing probabilities into Table 10.2. In the example, we judge that the *not-E & not-G* cell is the most probable: the most likely to represent the actual world. Let us say that $P(\textit{not-E \& not-G}) = .6$, and suppose further that $P(\textit{not-E \& G}) = 0$, as Carol has no reason to go to Glasgow when Arthur is not in Edinburgh. There is some probability that someone other than Carol could help Arthur go to Edinburgh, but it is far more likely that Carol would help him: $P(E \& G) = .1$ and $P(E \& \textit{not-G}) = .3$. Table 10.2 shows that *E & G* is the only cell in which *if E then G* is true on a Stalnaker analysis of the example, so $P(\textit{if E then G}) = P(E \& G) = .1$. But then the Equation fails to hold, with $P(G \mid E) = P(E \& G)/P(E) = .1/.4 = .25$. The underlying general problem is that $P(B \mid A)$ depends only on the *A* possibilities, but in the Stalnaker and Lewis systems, the *not-A* possibilities are relevant to $P(\textit{if A then B})$. At one extreme, *if A then B* can be true at all the *not-A* possibilities, and then $P(\textit{if A then B})$ can be higher than $P(B \mid A)$, and at the other extreme, *if A then B* can be false at all the *not-A* possibilities, and then $P(\textit{if A then B})$ can be lower than $P(B \mid A)$. In general, $P(\textit{if A then B})$ and $P(B \mid A)$ are almost never identical.

Stalnaker's analysis of conditionals has influenced psychologists, but there is a serious problem with taking it as the computational basis of a psychological account of conditionals (Evans and Over, 2004). Lewis's proof implies that Stalnaker's theory is in fact incompatible with the Equation, that $P(\textit{if A then B}) = P(B \mid A)$, but psychologists of reasoning have found strong support for this identity when it is interpreted as an empirical claim and studied experimentally as the *conditional probability hypothesis*: that $P(\textit{if A then B}) = P(B \mid A)$ in ordinary people's judgments.

The first experiments on this hypothesis gave the participants information about a frequency distribution of cards in an artificial pack, and the indicative conditional was about a card that was to be randomly selected from the pack (Evans et al., 2003; see also Oberauer and Wilhelm, 2003). For instance, the pack might contain four cards: a yellow card with a circle on it, a yellow card with a diamond, a red card with a circle, and a red card with a diamond. The conditional might be, *if the card is yellow (Y) then it has a circle on it (C)*. The participants would be told that this is a singular conditional about the specific card that will soon be randomly selected from the pack, and they would be asked for a probability judgment, *P(if Y then C)*. In experiments like this, participants can easily use the frequency distribution to make their probability judgments, and the findings are that most of them judge that the probability of the conditional is the conditional probability, and virtually none say that it is the probability of the material conditional. On this evidence, most people in our example would judge that $P(if\ Y\ then\ C) = P(C \mid Y) = .5$, and virtually none would judge that $P(if\ Y\ then\ C) = P(not\ Y\ or\ C) = .75$ (for further relevant experiments, see Cruz and Oberauer, 2014; Evans et al., 2007; Fugard et al., 2011; Oberauer et al., 2007).

Girotto and Johnson-Laird (2004) have attempted to explain away the empirical findings in support of the Equation as arising from a misinterpretation of the task by participants. Their basic argument is that, when participants are asked for the probability of a conditional, *if A then B*, they interpret the question as if it were about the probability of *B* if *A* holds, where the probability operator is taken to apply to the consequent and not to the conditional as a whole. But as modal logicians standardly point out (Garson, 2014), modal operators in English conditionals, like those for necessity and probability, can apply semantically to the whole conditional, even if they appear syntactically in the consequents. More technically, these operators create scope ambiguities in English, and Girotto and Johnson-Laird were unjustified in simply assuming that participants had narrow scope readings of the conditional, applying only to the consequent, and not wide scope readings, applying to the conditional as a whole (Over and Baratgin, 2017; Over et al., 2013; Over et al., 2007; Politzer et al., 2010).

Girotto and Johnson-Laird (2004) further assumed that a conditional probability judgment results from taking the probability operator to apply to the consequent in a conditional (Over and Baratgin, 2017). This is a modal fallacy that goes back to classical times and Aristotle's discussion of necessary truth, and whether what is going to happen will necessarily happen (Gaskin, 1995). For a simple example, it is trivial that, for all *A*, our confidence in *if A then A* (wide scope) is at the highest degree of 1, and this corresponds to the trivial conditional probability that $P(A \mid A) = 1$. In contrast, it is anything but trivial to claim that, for all *A*, if *A* holds, then our confidence (narrow scope) in *A* is at 1, for that is to claim omniscience. The Equation, *P(if A then B) = P(B | A)*, is about a wide scope reading of subjective probability, applied to the whole conditional (Edgington, 1995), and the experimental evidence so far is that people do have this wide scope interpretation (Over et al., 2013).

Further support for the conditional probability hypothesis comes from experiments on *causal conditionals*: conditionals justified by reference to a causal structure. Another background story could turn (1), *if E then G*, into a highly probable causal conditional. Let us say that Arthur is fit and well, and that he and Carol are academic collaborators who have just agreed that Arthur will present their joint research at a workshop in Edinburgh while Carol gives a talk on it at the same time in a conference at Glasgow. Now there is no relevant frequency information, for *P(if E then G)*, about how many times Carol has been in Glasgow when Arthur has been in Edinburgh. There are, however, possible disabling conditionals for *if E then G*, that could affect *P(if E then G)*. There might, for instance, be some chance of a train drivers' strike, which would prevent Carol from going to Glasgow no matter where Arthur is. Knowledge of the causal structure of examples like this, with its possible disabling conditions for the conditional, could be represented in a causal model that people construct, and this could be the basis of their judgments about *P(if E then G)* and of the inferences they make from *if E then G* (Ali et al., 2011; Fernback and Erb, 2013; Oaksford and Chater, 2013).

In experiments on causal conditionals, with possible disabling conditionals in the background, conditional probability is the best predictor of the probability of the conditional in participants' judgments (Over et al., 2007; Singmann et al., 2014), and there is again virtually no support for the material conditional analysis of these conditionals (further empirical support for the conditional probability response over a range of different conditionals can be found in Baratgin et al., 2013, Feeney and Handley, 2011, Haigh et al., 2013, Politzer et al., 2010, and van Wijnbergen-Huitink et al., 2015).

A possible qualification of the conditional probability hypothesis could come from studies of *missing-link conditionals*, defined as conditionals without any relation between the antecedent and consequent. Yet another background story could turn (1) into a missing-link conditional. Perhaps Arthur's going to Edinburgh would have no effect on whether Carol goes to Glasgow. Arthur and Carol do not know each other. Arthur lives in Australia, and Carol lives near Glasgow and often goes there on business that has nothing to do with Arthur. Then *P(E)* and *P(G)* are independent of each other, and *P(G | E)* is high simply because *P(G)* is high. There are arguments that *missing-link conditionals* like this should be rejected as false or improbable (Douven, 2016; Krzyżanowska et al.,

2013). The question is whether the Equation, $P(\text{if } E \text{ then } G) = P(G \mid E)$, holds for missing-link conditionals.

Skovgaard-Olsen et al. (2016) found most participants conform to the Equation only when a relation between A and B made $P(B|A) > P(B|not\text{-}A)$, implying that A raises the probability of B. Other results, however, do not support this conclusion and instead provide further evidence for the conditional probability hypothesis (Oberauer et al., 2007; Over et al., 2007; Singmann et al., 2014). A possible reason for these divergent findings is that people might sometimes interpret a question about the probability of a singular indicative conditional as about assessing causal power: the strength of a causal relation between A and B. With the latter interpretation, it becomes trivially true that judging how far A raises the probability of B is integral to the assessment (see Oberauer et al., 2007 for an explicit comparison of the effect of the two variables).

An interpretation of questions about conditionals as questions about causal power (whether from cause to effect or from effect to cause) seems relevant mostly for conditionals with causal content. This would help explain why researchers who hold that missing-link conditionals are false, or highly improbable, have restricted their claim to so-called inferential conditionals as opposed to "content conditionals" (Krzyżanowska et al., 2013), or to conditionals containing real-world materials as opposed to conditionals using pseudo-naturalistic or abstract materials, the latter being designed to reduce the influence of world knowledge, including about causal relations, on people's responses (Skovgaard-Olsen et al., 2016). A restriction on the scope of a theory would ideally be justified from within the theory rather than by empirical findings that the theory cannot account for. The observation that $P(B|A) > P(B|not\text{-}A)$ plays a role in people's judgments of the probability of conditionals in some experiments, but not in others, goes against the idea that it is at the semantic core of conditionals. This is also the case if the effect is not found to be specific to conditionals.

Consider a *missing-link disjunction*, A or B, with no relation at all between A and B. When Arthur and Carol have nothing to do with each other, the disjunction "Arthur is not in Edinburgh or Carol is in Glasgow" is arguably just as odd in some sense as (1) when there is no link between E and G. Cruz et al. (2016) have indeed found that the negative effect of a "missing link" is not specific to conditionals, but is also present in disjunctions and conjunctions. This finding suggests that the effect is pragmatic and results from the lack of a common topic between A and B, in *if A then B, A or B*, and *A & B*.

Further research on this topic could investigate an even wider range of missing-link cases. Suppose Carol promises Arthur that she will go to Glasgow if he goes to Edinburgh, when she intends to go to Glasgow (for reasons of her own) no matter what Arthur does. There is certainly something wrong with this promise, but is it semantic? Carol's misleading promise would seem to be worse, and more open to criticism, if Arthur has much trouble and expense going to Edinburgh, in the mistaken belief that this will cause Carol to go to Glasgow. Being misleading in an assertion, to someone else's cost, is intuitively a pragmatic fault and not a semantic one, but experimental evidence would be relevant to this claim (see also Douven et al., 2018). In any case, there is very strong empirical support for the Equation applied to standard uses of indicative conditionals, without missing links.

Psychologists have long known that people respond with what were initially called "defective" truth tables when they were asked about the truth value of indicative conditionals (Evans and Over, 2004; and see Over and Baratgin, 2017, for a recent review). For instance, Politzer et al. (2010) presented participants in an experiment with the following conditional about a chip that was to be selected at random from a distribution of seven chips, which were either round or square, and either black or white:

(2) If the chip is square (S), then it will be black (B).

The participants tended to respond that (2) was true when the chip turned out to be square and black, S and B, false when it was square and not black, S and *not-B*, and neither true nor false when it was not a square, *not-S*. On Stalnaker's analysis of the conditional, it is always true or false.

It could be argued, philosophically, that a Stalnaker-type analysis does not apply to conditionals *if A then B* with a false A, when it is impossible to determine the closest possible world in which A is true, to see whether or not B is true there. Politzer et al. (2010) do not give their participants any way to make such a determination, although it is possible to do that in more complex experimental designs (van Wijnbergen-Huitink et al., 2015). Nevertheless, Stalnaker's position as it stands cannot explain the data in experiments like that of Politzer et al., whose results instead support the Equation as general description of participants' judgments about the probability of an indicative conditional.

3 The de Finetti and Jeffrey tables

Philosophers and logicians, starting with Aristotle (Gaskin, 1995), have proposed a wide range of three-valued and many-valued logical systems that could be relevant to the psychology of reasoning (Baratgin et al., 2013; Gottwald, 2015), but the three-value system suggested by de Finetti (1936/1995, 1937/1964), for what he termed a conditional event, has had the most impact on psychological studies and theories of conditionals (Evans et al., 2015; Over and Baratgin, 2017; Pfeifer, 2013; Pfeifer and Kleiter, 2010; Politzer et al., 2010; Singmann et al., 2014). Both de Finetti and Ramsey (1929/1990) held that an indicative conditional is "void" when its antecedent is false, but Ramsey's position was only stated in a footnote, in what came to be known as the Ramsey test, and de Finetti was much more systematic. It is striking that the two founders of contemporary subjective probability theory classified an indicative conditional as "void" when its antecedent is false, and identified its probability with the conditional probability.

Table 10.3 summarizes what can be called *the de Finetti* table for (2), and it can be clearly related to the results of Politzer et al. (2010). When one is looking directly at a transparently round chip in that experiment, what can one justifiably say about the indicative conditional, "If the chip is square, then it is black," except that it is in some sense "void"? It would naturally be rephrased as a counterfactual, "If the chip had been square, then it would have been black." But its probability, *P(if S then B)*, in a de Finetti analysis, is the probability of the *S & B* outcome given that it has a (non-void) truth value, $P(S \& B \mid S)$, which is the same as $P(B \mid S)$, and so (2) is interpreted as a conditional event, and the Equation is satisfied (see also Baratgin et al., 2013).

In other experiments, in which the conditional has a more ordinary use, there are possible complications. Skovgaard-Olsen et al. (2017) study the example, "If Scott turns on the warm water, then Scott will be warm soon," with a scenario context in which Scott has been playing in snow. Clearly, there is a reason, even in a de Finetti approach, for applying "true" to this conditional in that context, for its conditional probability can be judged very high, if not 1, by the Ramsey test. Edgington (2003) referred to such cases as the *pleonastic* use of "true," and this use could also be called *pragmatic* (Adams, 1998). These uses of "true" are pragmatic endorsements of an assertion, and the word can be used in this way

Table 10.3 The de Finetti table for *if S then B*

B S	1	0
1	1	0
0	V	V

1 = true, 0 = false, and V = void.

for purely subjective judgments, for example, when one person says it is "true" that coq au vin should be made with red wine, while another (a lover of coq au Riesling) rejects that as "false" (Politzer et al., 2010). The possible existence of these uses, for both "true" and "false," makes truth-value gaps for conditionals hard to find in natural language contexts (Over et al., 2007).

Another possibility is that, when participants are told, for example, that Scott does not turn on the warm water, they give the indicative conditional a counterfactual interpretation, so that it becomes equivalent to, "If Scott were to turn on the warm water, then Scott would be warm soon." They could then state that the counterfactual is "true," because it is supported by a causal relationship. The problem is that participants will always try to make sense, in ordinary terms, of an evaluation or reasoning task that they are given, and it can take an extended series of experiments to tease apart different possibilities (Evans and Over, 2004).

The de Finetti table is certainly too limited to be an account of conditionals on its own, and in fact no one has ever proposed it without the Ramsey test as a supplement. Baratgin et al. (2013) point out that we clearly want to evaluate as "true" a logical truth like *if S & B then B*, and they refer to Jeffrey (1991), who showed that the "void" value in the de Finetti table for a conditional becomes, in a further differentiation of the de Finetti analysis, the conditional probability itself. More precisely, when A is false, the *expected value* of *if A then B* proves to be the conditional probability, $P(B \mid A)$. Most contemporary followers of de Finetti have agreed with Jeffrey (1991) in following this line (Coletti and Scozzafava, 2002; Pfeifer and Kleiter, 2009). The resulting table can be called the *Jeffrey table* (Cruz and Oberauer, 2014; Over, 2016, 2017; Over and Baratgin, 2017). See Table 10.4.

Consider the Jeffrey table for *if S & B then B*. This conditional will have value 1 in the first cell, in which S and B both have value 1. The second cell is impossible: S & B cannot have 1 there when B has 0. In the two cells in which S & B has value 0, the value of *if S & B then B* is 1, since $P(B \mid S \& B) = 1$. Consequently,

Table 10.4 The Jeffrey table for *if A then B*

B A	1	0
1	1	0
0	$P(B \mid A)$	$P(B \mid A)$

1 = true, 0 = false, and $P(B|A)$ = the subjective conditional probability of B given A.

the logical truth *if S & B then B* has value 1 in all the possibilities. What does 1 mean in the cells of this table? It could be argued that 1 refers to objective, factual truth when *S* and *B* both hold in the actual world. Supposing *S & B* is false in the actual world, the actual world does not make the nonmaterial conditional *if S & B then B* true. But the value of *if S & B then B* will be $P(B \mid S \& B) = 1$, because there is a logically necessary connection between *S & B* and *B*, and so *if S & B then B* has whatever kind of truth is indicated by logical truth. Here we are again back to philosophical questions that have been asked since Aristotle. A related psychological question is whether people have a univocal notion of truth, but people's use of "true" and "false" is a neglected topic in the psychology of reasoning. On the other hand, psychologists have extensively investigated reasoning with conditionals, and the new probabilistic approach to the subject illuminates this topic.

4 The probabilistic validity of inferences

Before turning to empirical findings on people's evaluation of inferences from uncertain premises, let us consider what it means for an inference from uncertain premises to be valid or invalid. The definition of validity in classical logic had no provision for uncertain beliefs—it specified only that an inference is valid when it would be inconsistent for its premises to be true and its conclusion false. In other words, an inference was valid when it preserved truth from premises to conclusion for all consistent truth-value assignments to the premises and the conclusion. Note that this definition of validity depends on the concept of logical consistency: there cannot be an assignment of both truth and falsity to any proposition *A*.

In the probabilistic approach to reasoning, consistency and validity have been generalized to allow the representation of uncertainty in the statements people reason about. Consistency is generalized to *coherence*: the probabilities of statements are coherent when they conform to the axioms of probability theory. For example, if we believe there is a .8 probability of rain today, then to be coherent we should also believe that there is a .2 probability of no rain today, for otherwise the numbers used to represent our degrees of belief would not sum to 1, violating an axiom of probability theory. When people's beliefs are incoherent in this sense, a Dutch book can be made against them (Vineberg, 2016).

Just as in binary, classical logic the definition of validity drew on the definition of consistency, the definition of *probabilistic validity*, *p-validity* for short, draws on the definition of coherence. A one-premise inference is p-valid when, for all coherent probability assignments to the premise and the conclusion, the probability of the conclusion is not lower than the probability of the premise. For the general case of inferences with any number of premises, let the uncertainty of a statement equal one minus its probability, $U(A) = 1 - P(A)$. Then an inference is p-valid when, for all coherent probability assignments to the premises and the conclusion, the uncertainty of the conclusion is not higher than the sum of the uncertainties of the premises (Adams, 1998; and see Adams, 1975, for an earlier definition of p-validity).

The condition of coherence in the definition of p-validity is important because it avoids a possible misunderstanding of p-validity. It might be thought that there are cases in which p-validity and coherence conflict, because a conclusion drawn for an inference could apparently satisfy the normative constraints of p-validity and yet be incoherent. For example, consider the trivial p-valid inference *A, therefore A*. If we assign a probability of .8 to the premise, then to be coherent we should also assign a probability of .8 to the conclusion. Assume we now apply the above definition of p-validity without taking into account that it refers to probability functions, which are strictly defined as conforming to the axioms of probability theory. For example, in the analysis of an experiment on the inference *A, therefore A*, we might try to claim that participants conform to p-validity, but not to coherence, when they assign a probability of .8 to *A* as the premise and .9 to *A* as the conclusion. However, this would be to ignore the fact that a probability function for defining p-validity must be coherent: there is no coherent probability assignment of both .8 and .9 to *A*. For this reason, there can be no conflict between coherence and p-validity.

Adams (1996) outlined different ways in which probability can be preserved from premises to conclusion, but he singled out p-validity as special because it coincides with classical validity in all cases in which an inference can be represented in both logical systems. Thus, all inferences that are classically valid are p-valid, and all inferences that are classically invalid are p-invalid. For example, it is both classical valid and p-valid to infer *not-A or B* from *not-A*, as $P(not\text{-}A \text{ or } B)$ cannot be coherently lower than $P(not\text{-}A)$. But the probability conditional / conditional event, defined in terms of the Equation, $P(if\ A\ then\ B) = P(B \mid A)$, cannot be represented in classical logic with its restriction to the two possibilities of truth and falsity. The inferences that are p-valid for the probability conditional are a proper subset of the corresponding inferences that are classically valid for the material conditional (Adams, 1998). For instance, it is not p-valid to infer *if A then B* from *not-A* for the probability conditional, since $P(B \mid A)$ can be coherently less than $P(not\text{-}A)$. The logic for the probability conditional has come to be known as *System P*, which can be seen as the logic of probability, or of partial belief, for this conditional (Pfeifer and Kleiter, 2009).

5 The psychology of conditional reasoning

Traditional psychology of reasoning primarily investigated people's inferences from assumptions, which could be abstract, arbitrary, or unbelievable. For example, participants might be asked to assume that no nutritional things are inexpensive, and that some vitamin tablets are inexpensive. They would then be asked whether it necessarily followed that some vitamin tablets are not nutritional. If any of them disagreed, they were classified as a having a *belief bias* (Evans and Over, 2004). They were rejecting a conclusion because it conflicted with their beliefs, even though it did follow validly from the assumptions. Belief bias is an important topic to investigate, but the traditional approach was severely limited because it did not focus on inferences from uncertain premises, including most of people's actual beliefs, which are not merely assumed to be true.

In the new Bayesian / probabilistic approach to the study of reasoning, we do not usually ask participants in experiments what follows from arbitrary assumptions that we give them. We often ask for how much confidence they have in conclusions given degrees of belief in the premises of inferences. Consider again or-introduction, inferring *A or B* from *A*, which is p-valid. As we have reported above, Johnson-Laird and Byrne (1991) claimed that people do not endorse its validity. Our reply was that people will not violate its validity in their degrees of belief, and Cruz et al. (2017) found that people do conform to the p-validity of or-introduction, by judging that *P(A or B)* is greater than or equal to *P(A)*, when they are asked for their subjective probability judgments about *A* and *A or B* (see also Politzer and Baratgin, 2016).

Most effective reasoning, whether in ordinary affairs or technical subjects and the sciences, is from premises that are believable or plausible to some extent, though not certain. Making arbitrary assumptions, which are additionally unbelievable, such as that no nutritional things are inexpensive, has little if any role to play in most human reasoning. If an assumption or supposition is made, in a Ramsey test, or at the start of a reductio ad absurdum argument, its purpose is usually to infer a believable conditional, or to derive a contradiction from an unbelievable premise, in the hope of inferring the negation of the premise, which will be believable. The recognition of these points has led to the development of a broad *new paradigm* in the psychology of reasoning, which is Bayesian or probabilistic (Elqayam and Over, 2013; Hahn and Oaksford, 2007; Oaksford and Chater, 2007; Over, 2009, 2016). This development brings with it a new logic for the natural language conditional, System P, and this fundamental change in logic justifies calling it a new paradigm.

Consider an example, based on (1), of the classical conditional inference, *Modus Ponens* (MP):

If Arthur is in Edinburgh (*E*), then Carol is in Glasgow (*G*).
Arthur is in Edinburgh.
Therefore, Carol is in Glasgow.

We could simply assume the major premise, *if E then G*, and the minor premise, *E*, and then of course the conclusion, *G*, would necessarily follow, under these assumptions. MP is a valid inference for all the accounts of conditionals we have covered. It is valid for the material conditional, and for the conditionals of Stalnaker (1968) and Lewis (1973). It is also valid for the probability conditional / conditional event. When $P(G \mid E)$ is assumed to be 1, and $P(E)$ to be 1, then $P(G)$ must also be 1. However, after first introducing (1) above, we discussed an example in which Arthur had broken legs and could not go to Edinburgh without Carol's help. In this case, we would not bother wasting energy on MP. Its major premise *if E then G* would be probably false, and we could have no confidence in its conclusion as a rational belief. But in another example we referred to, Arthur is fit and well and has a business agreement with Carol that makes (1) highly probable. Then with the minor premise *E* highly probable, we would use MP to extend our high confidence to the conclusion *G*.

We can analyze the two examples that we have just used more fully and precisely by applying the Equation and the axioms of subjective probability theory (Coletti and Scozzafava, 2002; Howson and Urbach, 2006; Pfeifer and Kleiter, 2009, 2010). The total probability theorem implies that

$$P(G) = P(E)P(G \mid E) + P(\text{not-}E)P(G \mid \text{not-}E)$$

By the Equation, the probability of the major premise of our example, *P(if E then G)*, is $P(G \mid E)$. In the first version of the example, in which Arthur has broken legs, let us now say that $P(\text{if } E \text{ then } G) = P(G \mid E) = 0$, because the only possible way that Arthur could get to Edinburgh would be if Carol went along to help him. We might also decide that $P(E)$ is extremely close to 0, making $P(\text{not-}E)$ extremely close to 1 by probability theory, because Arthur has no intention of travelling anywhere when he has broken legs. Even with this version of the example fleshed out this far, we may still be unable to make a judgment about $P(G \mid \text{not-}E)$. But its minimum value will be 0, and its maximum value will be 1, by probability theory. If $P(G \mid \text{not-}E) = 0$, $P(G) = 0$ by the total probability theorem, and if $P(G \mid \text{not-}E) = 1$, $P(G)$ is extremely close to 1, again by the theorem. Hence to be coherent with probability theory, $P(G)$ must lie in the interval between 0 and a value extremely close to 1, and clearly employing MP as an inference in this case would tell us almost nothing about $P(G)$ and would be a waste of time and energy.

We have already seen that, with $P(G \mid E) = P(E) = 1$, applying MP will give us $P(G) = 1$, which is useful. But supposing $P(G \mid E)$ and $P(E)$ are both uncertain and yet high, MP can still be useful for us. In the version of the example in which Arthur is fit and well, and he and Carol have the business agreement, it might be, say, that $P(G \mid E) = .8$ and $P(E) = .9$. Then by the total probability theorem, the minimum value of $P(G)$ will be .72, and the maximum value will be .82. As long as our inferred judgment about $P(G)$ is in this interval, between .72 and .82, we will be coherent by probability theory and will have a rational degree of belief, in this sense.

Psychologists have established results about what they have called the *suppression* of valid inferences (Byrne, 1989). One way to do this would be to add an additional premise to the instance of MP, for example, we could introduce a second conditional premise stating that, if Arthur has not broken his legs, then Carol is in Glasgow. Participants in an experiment could then lose confidence that the conclusion followed. Evidence has accumulated that suppression is the result of induced uncertainty in the premises. This can be caused by an additional premise (Politzer, 2005; Stevenson and Over, 1995), or simply by giving the premises, not as assumptions, but as more or less plausible statements in a dialogue (Stevenson and Over, 1995, 2001). There is an important distinction here that has been missed in most of the literature. We should distinguish between the suppression of a valid *inference* and the suppression of its *conclusion*. In a Bayesian / probabilistic study of inferences from uncertain premises, confidence in the conclusion of even a valid inference should be relatively low when the premises are relatively uncertain, but that does not necessarily mean that the inference itself has been suppressed (Over and Cruz, 2018).

Consider an instance of MP, *if E then G, E, therefore G*, in which $P(G \mid E) = .8$ and $P(E) = .9$. As we have just seen above, $P(G)$ should be between .72 and .82 for coherence. Suppose first that some people judge that $P(G)$ is .75 under these uncertain premises. In this case, we could rightly speak of the suppression of the conclusion if these people had more confidence in G before they made this inference. However, it would not be right to call this an instance of suppressing MP as an inference, for these people are coherent in their judgment about $P(G)$ under these premises. They could indeed be said to be rationally uncertain. The case would be completely different if they judged that $P(G)$ is .60. In this second case, they would be incoherent, and the MP inference would itself be suppressed. The probabilistic logic of MP would have been violated, with $P(G \mid E) = .8$ and $P(E) = .9$. This second case is one of irrationally low confidence.

It is important to note that, in a Bayesian analysis, people can also be irrationally overconfident in the conclusion of MP. In the example we have just considered, in which $P(G \mid E) = .8$ and $P(E) = .9$, people would be irrationally overconfident if they judge that $P(G)$ was .98. In the traditional, binary paradigm, the validity of MP was endorsed at ceiling, and the possibility of overconfidence in MP conclusions could not even be conceived, or expressed, with only the two possibilities of truth and falsity. Such overconfidence has not been found in the new paradigm as yet, but the question is far from being fully investigated.

6 Counterfactuals

Counterfactual reasoning is another important topic that has not yet been fully studied in the new paradigm. Consider this example from Over et al. (2007):

(3) If the cost of petrol had increased (*C*), then traffic congestion would have declined (*D*).

Counterfactuals like (3) have been extensively discussed and investigated in psychology from Bayesian and other perspectives (Over, 2017). Over et al. (2007) found that people's judgments about these counterfactuals do satisfy the Equation, $P(if\ C\ then\ D) = P(D|C)$. However, psychologists of reasoning have paid little attention to logical inferences for counterfactuals; even the study of MP for counterfactuals has received little attention (with the exception of Thompson and Byrne, 2002).

Experiments on MP for counterfactuals can help answer a fundamental question about them, for philosophical logicians and psychologists of reasoning. Does a counterfactual like (3), *if C then D*, logically imply that its antecedent is false, *not-C*? Clearly, (3) is in some way incompatible with the truth of its antecedent, *C*, but is this semantic, because *if C then D* logically implies *not-C*, or pragmatic, because speakers' use of *if C then D* pragmatically presupposes, or implicates in the sense of Grice (1989), that they believe *not-C* to a high degree? In the most influential philosophical accounts of counterfactuals, by Stalnaker

(1968) and Lewis (1973), *if C then D* does not logically imply *not-C*, and MP with a counterfactual major premise is a logically valid inference with consistent premises.

Lewis (1973) supported this position philosophically with a thought experiment about a dialogue. For instance, there could be two speakers talking about a foreign country that the second speaker knows better than the first. Suppose the first speaker asserts (3) about what could have happened in that country over the last year. It would not appear unnatural for the second speaker to assert the minor premise for MP: that in fact the cost of petrol did increase in the country over the past year. It would then also seem to be legitimate for both speakers to use MP to infer that traffic congestion had declined there. The second speaker would not be contradicting the semantic content of the first speaker's conditional, but would be cancelling a pragmatic presupposition or implicature (Grice, 1989). Thompson and Byrne (2002) ran properly designed and controlled experiments on dialogue examples like this, and found that the participants did not react as if the second speaker had contradicted the first, so that the conditional content could not be trusted, but were, rather, happy to endorse the MP conclusion. This result supports the hypothesis that a counterfactual does not logically imply that its antecedent is false, but that the falsity of the antecedent is a pragmatic aspect of a speaker's use of a counterfactual, which can be fairly easily cancelled.

There has as yet been little advance in experiments on logical inferences, like MP, for counterfactuals beyond Thompson and Byrne (2002). A fundamental inference that they did not study is *centering*. There is *one-premise centering*, which is inferring *if A then B* from *A & B*, and two-premise centering, which is inferring *if A then B* from *A* and *B* as separate premises. (One-premise centering is often called "conjunctive sufficiency"; see Hajek and Hall, 1994, on our use of centering.) Centering (both one-premise and two-premise) separates different theories of the conditional at a high level. It is valid for the material conditional, the Stalnaker / Lewis conditionals, and the probability conditional. It is invalid in theories in which the actual world is not uniquely the closest possibility, with the conditional only holding when the consequent is true in a set of possibilities (including the actual world) in which the antecedent is true (Cariani and Rips, 2017; Kratzer, 2012). It is also invalid for conditionals that only hold when there is a connection between the antecedent and the consequent (as in the accounts of Krzyżanowska et al., 2013; Skovgaard-Olsen et al., 2016). Centering is relevant to the question of how counterfactuals are related to causation, which is of great interest to both philosophers and psychologists (Edgington, 2011; Hoerl et al., 2011; Menzies, 2014; Over; 2017). For example, how is (3) related to the possibility that increasing the cost of petrol causes traffic congestion to decline? It is hard to argue that the relation is especially close if (3) can be validly inferred merely from an increase in the cost of petrol and a decline in traffic congestion.

There is already experimental evidence that people endorse centering for the indicative conditional (Cruz et al., 2015, 2016; Politzer and Baratgin, 2016). Studies of centering and suppression, which we referred to above, could be extended to counterfactuals using the dialogue technique. There is also the possibility of demonstrating the *enhancement* of confidence in a counterfactual premise and so in the conclusion of MP (and other inferences) from that premise.

Enhancement could be studied for indicative conditionals, but consider the classic examples of Adams (1970):

(4) If Oswald did not kill Kennedy, then someone else did.
(5) If Oswald had not killed Kennedy, then someone else would have.

Adams used (4) and (5) to demonstrate how strikingly indicative and counterfactual conditionals can differ from each other. For most of us, (4) is probably true, and (5) is probably false. But suppose that, after we doubt (5), we read the latest findings of a scientific study, with conclusive evidence that Oswald did not kill Kennedy. At that point, our confidence in (5) would be enhanced, and we would have high confidence in the MP conclusion that someone else killed Kennedy (Over et al., 2017).

More generally, enhancement and suppression are cases in which we change our degrees of belief in light of new information, and our conditional probability judgments are not *rigid* or *invariant*, but change as well. Belief revision / updating, with and without invariance in the conditional probabilities, is a central topic in the new paradigm, which we do not have the space here to cover (Oaksford and Chater, 2013).

Psychologists have not only been interested in the relation between counterfactuals and causal judgments, but also in their connection to deontic judgments and expressions of emotion, particularly regret (Byrne, 2016). Politicians might consider the deontic conditional, "If traffic congestion is to decline, then we must increase the tax on petrol." Suppose that, after a debate, they do not increase the petrol tax, but traffic congestion remains a serious problem, and deaths from traffic accidents increase as a result. Reflecting later on the probable truth of (3), the politicians could feel regret, especially if they had been "close" to increasing the tax on petrol (Kahneman and Miller, 1986). There is much of potential interest to philosophers in this psychological research, and in other psychological studies of deontic and utility conditionals and related inferences. Unfortunately, we have now run out of space to cover these topics (but see Bonnefon and Sloman, 2013; Elqayam et al., 2015; Over and Over, 2017).

7 Conclusions

Traditional psychology of reasoning focused primarily on inferences from given assumptions, and the most influential theory of the natural language conditional in this tradition (Johnson-Laird and Byrne, 1991) made a false start, we have argued, by representing this conditional, when fully understood, as essentially the truth functional material conditional of classical logic. It would have been better to take account of earlier logical analyses of this conditional that were not truth functional (Stalnaker, 1968; Lewis, 1973). More recent developments in the psychology of reasoning have been Bayesian / probabilistic, and grounded in formal and philosophical research on reasoning, probability, conditionals going back to de Finetti (1936/1995) and Ramsey (1926/1990; 1929/1990), and what they called the logic of probability and of partial belief. This new approach focuses on reasoning from uncertain beliefs, and on how beliefs change over time. It will continue to benefit from close cooperation between philosophers and psychologists especially as it extends its attention to reasoning with counterfactuals.

References

Adams, E. (1970). Subjunctive and indicative conditionals. *Foundations of Language*, 6:89–94.

Adams, E. (1975). *The Logic of Conditionals: An Application of Probability to Deductive Logic.* Dordrecht, The Netherlands: Reidel.

Adams, E. (1996). Four probability-preserving properties of inferences. *Journal of Philosophical Logic*, 25(1):1–24.

Adams, E. (1998). *A Primer of Probability Logic.* Stanford, CA: CLSI Publications.

Ali, N., Chater, N. and Oaksford, M. (2011). The mental representation of causal conditional reasoning: Mental models or causal models. *Cognition*, 119:403–18.

Baratgin, J., Douven, I., Evans, J. St. B. T., Oaksford, M., Over, D. E. and Politzer, G. (2015). The new paradigm and mental models. *Trends in Cognitive Sciences*, 19(10):547–48.

Baratgin, J., Over, D. E. and Politzer, G. (2013). Uncertainty and de Finetti tables. *Thinking & Reasoning*, 19:308–28.

Bonnefon, J. F. and Sloman, S. A. (2013). The causal structure of utility conditionals. *Cognitive Science*, 37:193–209.

Byrne, R. M. J. (1989). Suppressing valid inferences with conditionals. *Cognition*, 31:61–83.

Byrne, R. M. J. (2016). Counterfactual thought. *Annual Review of Psychology*, 67:135–57.

Cariani, F. and Rips, L. J. (2017). Conditionals, context, and the suppression effect. *Cognitive Science*, 41:540–89.

Coletti, G. and Scozzafava, R. (2002). *Probabilistic Logic in a Coherent Setting*. Dordrecht, The Netherlands: Kluwer.

Cruz, N., Baratgin, J., Oaksford, M. and Over, D. E. (2015). Bayesian reasoning with ifs and ands and ors. *Frontiers in Psychology*, 6:192.

Cruz, N. and Oberauer, K. (2014). Comparing the meanings of "if" and "all". *Memory & Cognition*, 42:1345–56.

Cruz, N., Over, D., Oaksford, M. and Baratgin, J. (2016). Centering and the meaning of conditionals. In A. Papafragou, D. Grodner, D. Mirman and J. C. Trueswell (Eds.), *Proceedings of the 38th Annual Conference of the Cognitive Science Society*, 1104–09. Austin, TX: Cognitive Science Society.

Cruz, N., Over, D., & Oaksford, M. (2017). The elusive oddness of or-introduction. In G. Gunzelmann, A. Howes, T. Tenbrink, & E. Davelaar (Eds.). The *39th Annual Meeting of the Cognitive Science Society*, pp. 259–64. London: Cognitive Science Society.

de Finetti, B. (1936/1995). The logic of probability. *Philosophical Studies*, 77:181–90.

de Finetti, B. (1937/1964). Foresight: its logical laws, its subjective sources. In H. E. Kyburg and H. E. Smokier (Eds.), *Studies in Subjective Probability*, 55–118. New York, NY: Wiley.

Douven, I. (2016). *The Epistemology of Indicative Conditionals*. Cambridge: Cambridge University Press.

Douven, I., Elqayam, S., Singmann, H. and van Wijnbergen-Huitink, J. (2018). Conditionals and inferential connections: A hypothetical inferential theory. *Cognitive Psychology*, 101:50–81.

Edgington, D. (1995). On conditionals. *Mind*, 104:235–329.

Edgington, D. (2003). What if? Questions about conditionals. *Mind and Language*, 18:380–401.

Edgington, D. (2011) Causation first: Why causation is prior to counterfactuals. In C. Hoerl, T. McCormack and S. R. Beck (Eds.), *Understanding Counterfactuals, Understanding Causation: Issues in Philosophy and Psychology*, 230–41. Oxford: Oxford University Press.

Edgington, D. (2014). Indicative conditionals. In Edward N. Zalta (Ed.), *The Stanford Encyclopedia of Philosophy*. Retrieved from http://plato.stanford.edu/archives/win20 14/entries/conditionals/

Elqayam, S. and Over, D. E. (2013). New paradigm psychology of reasoning: An introduction to the special issue edited by S. Elqayam, J. F. Bonnefon and D. E. Over. *Thinking & Reasoning*, 19:249–65.

Elqayam, S., Thompson, V., Wilkinson, M., Evans, J. St. B. T. and Over, D. E. (2015). Deontic introduction: A theory of inference from is to ought. *Journal of Experimental Psychology: Learning, Memory, and Cognition*, 41(5):1516–32.

Evans, J. St. B. T., Handley, S. J., Neilens, H. and Over, D. E. (2007). Thinking about conditionals: A study of individual differences. *Memory & Cognition*, 35(7):1772–84.

Evans, J. St. B. T., Handley, S. J. and Over, D. E. (2003). Conditional and conditional probability. *Journal of Experimental Psychology: Learning, Memory, and Cognition*, 29:321–35.

Evans, J. St. B. T. and Over, D. E. (2004). *If*. Oxford: Oxford University Press.

Evans, J. St. B. T., Thompson, V. and Over, D. E. (2015). Uncertain deduction and conditional reasoning. *Frontiers in Psychology*, 6:398.

Feeney, A. and Handley, S. (2011). Suppositions, conditionals, and causal claims. In C. Hoerl, T. McCormack and S. R. Beck (Eds.), *Understanding Counterfactuals, Understanding Causation*, 242–62. Oxford: Oxford University Press.

Fernbach, P. M. and Erb, C. D. (2013). A quantitative model of causal reasoning. *Journal of Experimental Psychology: Learning, Memory, and Cognition*, 39:1327–43.

Fugard, J. B., Pfeifer, N., Mayerhofer, B. and Kleiter, G. D. (2011). How people interpret conditionals: Shifts toward conditional event. *Journal of Experimental Psychology: Learning Memory and Cognition*, 37:635–48.

Garson, J. (2014). Modal logic. In Edward N. Zalta (Ed.), *The Stanford Encyclopedia of Philosophy*. Retrieved from http://plato.stanford.edu/archives/sum2014/entries/logic-modal/

Gaskin, R. (1995). *The Sea Band the Master Argument: Aristotle and Diodorus Cronus on the Metaphysics of the Future*. Berlin: Walter de Gruyter.

Gilio, A. and Over, D. E. (2012). The psychology of inferring conditionals from disjunctions: A probabilistic study. *Journal of Mathematical Psychology*, 56:118–31.

Girotto, V. and Johnson-Laird, P. N. (2004). The probability of conditionals. *Psychologia*, 47:207–25.

Gottwald, S. (2015). Many-valued logic. In Edward N. Zalta (Ed.), *The Stanford Encyclopedia of Philosophy*. Retrieved from http://plato.stanford.edu/archives/spr2015/entries/logic-many-valued/

Grice, P. (1989). *Studies in the Way of Words*. Cambridge, MA: Harvard University Press.

Hahn, U. and Oaksford, M. (2007). The rationality of informal argumentation: A Bayesian approach to reasoning fallacies. *Psychological Review*, 114:646–78.

Haigh, M., Stewart, A. J. and Connell, L. (2013). Reasoning as we read: Establishing the probability of causal conditionals. *Memory & Cognition*, 41:152–58.

Hajek, A. and Hall, N. (1994). The hypothesis of the conditional construal of conditional probability. In E. Eells and B. Skyrms (Eds.), *Probability and Conditionals*, 75–111. Cambridge: Cambridge University Press.

Hoerl, C., McCormack, T. and Beck, S. (Eds.) (2011). *Understanding Causation, Understanding Counterfactuals*. Oxford: Oxford University Press.

Howson, C. and Urbach, P. (2006). *Scientific Reasoning: The Bayesian Approach.* Peru, NE: Open Court.

Jeffrey, R. C. (1991). Matter of fact conditionals. *Aristotelian Society Supplementary Volume,* 65:161–83.

Johnson-Laird, P. N. and Byrne, R. M. J. (1991). *Deduction.* Hove & London: Erlbaum.

Johnson-Laird, P. N., Khemlani, S. and Goodwin, G. P. (2015). Logic, probability, and human reasoning. *Trends in Cognitive Science,* 19:201–14.

Kahneman, D. and Miller, D. (1986). Norm theory: Comparing reality to its alternatives. *Psychological Review,* 93:136–56.

Kneale, W. and Kneale, M. (1962). *The Development of Logic.* Oxford: Oxford University Press.

Kratzer, A. (2012). *Modals and Conditionals.* Oxford: Oxford University Press.

Krzyżanowska, K., Wenmackers, S. and Douven, I. (2013). Inferential conditionals and evidentiality. *Journal of Logic, Language and Information,* 22(3):315–34.

Lewis, D. (1973). *Counterfactuals.* Cambridge, MA: Harvard University Press.

Lewis, D. (1976). Probabilities of conditionals and conditional probabilities. *Philosophical Review,* 85:297–315.

Menzies, P. (2014). Counterfactual theories of causation. In Edward N. Zalta (Ed.), *The Stanford Encyclopedia of Philosophy.* Retrieved from http://plato.stanford.edu/arch ives/spr2014/entries/causation-counterfactual/

Oaksford, M. and Chater, N. (2007). *Bayesian Rationality: The Probabilistic Approach to Human Reasoning.* Oxford: Oxford University Press.

Oaksford, M. and Chater, N. (2013). Dynamic inference and everyday conditional reasoning in the new paradigm. *Thinking & Reasoning,* 19:346–79.

Oberauer, K., Geiger, S. M., Fischer, K. and Weidenfeld, A. (2007). Two meanings of "if"? Individual differences in the interpretation of conditionals. *The Quarterly Journal of Experimental Psychology,* 60(6):790–819.

Oberauer, K., Weidenfeld, A. and Fischer, K. (2007). What makes us believe a conditional? *Thinking & Reasoning,* 13(4):340–69.

Oberauer, K. and Wilhelm, O. (2003). The meaning(s) of conditionals: Conditional probabilities, mental models, and personal utilities. *Journal of Experimental Psychology: Learning, Memory, and Cognition,* 29:680–93.

Orenes, I. and Johnson-Laird, P. N. (2012). Logic, models, and paradoxical inferences. *Mind & Language,* 27(4):357–77.

Over, D. E. (2009). New paradigm psychology of reasoning. *Thinking & Reasoning,* 15:431–38.

Over, D. E. (2016). The paradigm shift in the psychology of reasoning: The debate. In L. Macchi, M. Bagassi and R. Viale (Eds.), *Cognitive Unconscious and Human Rationality,* 79–97. Cambridge, MA: MIT Press.

Over, D. E. (2017). Causation and the probability of causal conditionals. In M. Waldmann (Ed.), *The Oxford Handbook of Causal Reasoning,* 307–25. Oxford: Oxford University Press.

Over, D. E. and Baratgin, J. (2017). The "defective" truth table: Its past, present, and future. In N. Galbraith, D. E. Over and E. Lucas, (Eds.), *The Thinking Mind: The Use of Thinking in Everyday Life*, 15–28. Hove: Psychology Press.

Over, D. E. and Cruz, N. (2018). Probabilistic accounts of conditional reasoning. In Linden J. Ball and Valerie A. Thompson (Eds.), *International Handbook of Thinking and Reasoning*, 434–50. Hove: Psychology Press.

Over, D. E., Cruz, N. and Oaksford, M. (2017). Dynamic reasoning with counterfactuals: Extending the Thompson-Byrne technique. In *Talk presented at the London Reasoning Workshop*, London, July 25, 2017.

Over, D. E., Douven, I. and Verbrugge, S. (2013). Scope ambiguities and conditionals. *Thinking & Reasoning*, 19:284–307.

Over, D. E., Hadjichristidis, C., Evans, J. St. B. T., Handley, S. J. and Sloman, S. A. (2007). The probability of causal conditionals. *Cognitive Psychology*, 54:62–97.

Over, H. and Over, D. E. (2017). Deontic reasoning and social norms: Broader implications. In N. Galbraith, D. E. Over and E. Lucas, (Eds.), *The Thinking Mind: The Use of Thinking in Everyday Life*, 54–65. Hove: Psychology Press.

Pfeifer, N. (2013). The new psychology of reasoning: A mental probability logical perspective. *Thinking & Reasoning*, 19:329–45.

Pfeifer, N. and Kleiter, G. D. (2009). Framing human inference by coherence based probability logic. *Journal of Applied Logic*, 7:206–17.

Pfeifer, N. and Kleiter, G. D. (2010). The conditional in mental probability logic. In M. Oaksford and N. Chater (Eds.), *Cognition and Conditionals: Probability and Logic in Human Thinking*, 153–73. Oxford: Oxford University Press.

Politzer, G. (2005). Uncertainty and the suppression of inferences. *Thinking & Reasoning*, 11:5–33.

Politzer, G. and Baratgin, J. (2016). Deductive schemas with uncertain premises using qualitative probability expressions. *Thinking & Reasoning*, 22:78–98.

Politzer, G., Over, D. E. and Baratgin, J. (2010). Betting on conditionals. *Thinking & Reasoning*, 16:172–97.

Ramsey, F. P. (1926/1990). Truth and probability. In D. H. Mellor (Ed.), *Philosophical Papers*, 52–94. Cambridge: Cambridge University Press.

Ramsey, F. P. (1929/1990). General propositions and causality. In D. H. Mellor (Ed.), *Philosophical Papers*, 145–63. Cambridge: Cambridge University Press.

Rips, L. J. (1983). Cognitive processes in propositional reasoning. *Psychological Review*, 90(1):38–71.

Singmann, H., Klauer, K. C. and Over, D. E. (2014). New normative standards of conditional reasoning and the dual-source model. *Frontiers in Psychology*, 5:316.

Skovgaard-Olsen, N., Kellen, D., Krahl, H. and Klauer, C. (2017). Relevance differently affects the truth, acceptability, and probability evaluations of "and", "but", "therefore", and "if then". *Thinking & Reasoning*, 23:449–82

Skovgaard-Olsen, N., Singmann, H. and Klauer, K. C. (2016). The relevance effect and conditionals. *Cognition*, 150:26–36.

Stalnaker, R. (1968). A theory of conditionals. In N. Rescher (Ed.), *Studies in Logical Theory*, 98–112. Oxford: Blackwell.

Stalnaker, R. (1970). Probability and conditionals. *Philosophy of Science*, 37:64–80.

Stevenson, R. J. and Over, D. E. (1995). Deduction from uncertain premises. *Quarterly Journal of Experimental Psychology*, 48A:613–43.

Stevenson, R. J. and Over, D. E. (2001). Reasoning from uncertain premises: Effects of expertise and conversational context. *Thinking & Reasoning*, 7:367–90.

Thompson, V. A. and Byrne, R. M. J. (2002). Reasoning counterfactually: Making inferences about things that didn't happen. *Journal of Experimental Psychology: Learning, Memory, and Cognition*, 28:1154–70.

van Wijnbergen-Huitink, J., Elqayam, S. and Over, D. E. (2015). The probability of iterated conditionals. *Cognitive Science*, 39:788–803.

Vineberg, S. (2016). Dutch book arguments. In Edward N. Zalta (Ed.), *The Stanford Encyclopedia of Philosophy*. Retrieved from https://plato.stanford.edu/archives/spr2016/entries/dutch-book/

Folk Judgments about Conditional Excluded Middle

Michael J. Shaffer and James R. Beebe

In this chapter we consider three philosophical perspectives (including those of Stalnaker and Lewis) on the question of whether and how the principle of conditional excluded middle should figure in the logic and semantics of counterfactuals. We articulate and defend a third view that is based upon belief revision theories in the tradition of the Ramsey Test. Unlike Lewis's view, the belief revision perspective does not reject conditional excluded middle, and unlike Stalnaker's, it does not embrace supervaluationism. We adduce both theoretical and empirical considerations to argue that the belief revision perspective should be preferred to Stalnaker's and Lewis's views. The empirical considerations are drawn from the results of three empirical studies ($N = 525$) of nonexperts' judgments about counterfactuals and conditional excluded middle.

1 Introduction[1]

At least since W. V. Quine introduced the Bizet/Verdi case in 1950 there has been significant controversy not only about the possibility of there being any adequate analysis of the logic of counterfactual conditionals, but also more specifically about the acceptability of the principle known as conditional excluded middle (CEM) (Quine, 1950). Conditional excluded middle is usually parsed as follows:

$$(CEM)(A > C) \vee (A >\sim C).$$

In other words, for any pair of conditionals with a common antecedent and whose consequents are a statement and its negation, at least one of the conditionals must be true. CEM is a consequence of what Daniel Bonevac calls "Stalnaker's rule" (Bonevac, 2003). This is stated as follows:

$$(SR) \frac{\sim (A > C)}{A > \sim C}$$

The acceptability of CEM was a particular bone of contention between Robert Stalnaker and David Lewis in developing their respective accounts of the logic and semantics of counterfactuals in the late 1960s and 1970s. Stalnaker ultimately argued that this principle should be incorporated in the logic of counterfactuals, and in so doing he favored the conditional logic C2. On this basis he argued further that we must introduce vagueness into the semantics for such conditionals (Stalnaker, 1981). In point of fact, he advocated doing this via the use of the theory of supervaluations that had previously been developed by Bas van Fraassen (Van Fraassen, 1966). The result is a semantic theory that allows conditionals in Stalnaker's logic to be true, false, or indeterminate. Lewis also acknowledged the need for vagueness in the semantics for counterfactuals, but he rejected CEM (Lewis, 1973).[2]

The main reasons why Stalnaker advocated this approach to the semantics of counterfactuals are twofold. First, it is supposed to explain our inability to choose a unique and most acceptable conditional from among competing conditionals like those in the Bizet/Verdi case. Second, it supports Stalnaker's personal conviction that CEM is a plausible principle for conditional logic. However, as we shall see, our inability to choose a unique, most epistemically acceptable conditional from among competing conditionals in Bizet/Verdi cases can be better explained without recourse to a semantics that incorporates vagueness and that we should not attempt to settle the issue of how to deal with such cases on the basis of intuitions about CEM (Shaffer, 2016). In this chapter, we test Stalnaker's and Lewis's theories against an alternative theory that also explains the inability to choose between competing conditionals on the basis of purely epistemic considerations. In addition to adducing theoretical reasons in support of the epistemic alternative, we report the results of three mixed method studies of folk judgments about counterfactuals and CEM in order to see which theory best fits with lay intuitions.

This debate importantly arose in virtue of the following pair of conditionals that Quine famously discussed in his 1950 book:

(BV1) If Bizet and Verdi had been compatriots, Bizet would have been Italian.
(BV2) If Bizet and Verdi had been compatriots, Verdi would have been French.

What this pair of conditionals is supposed to show is that there can be ties in terms of the closeness of non-actual possible worlds and so Stalnaker's analysis of the logic of counterfactuals is supposed to fail. The basic idea here is that there is good reason to suppose that worlds where Bizet and Verdi are both French or are both Italian are more similar to the actual world than worlds where they are, for example, Nigerian, Australian, or Sri Lankan. Yet it seems to be the case that there is no good reason to suppose either that the world where they are both Italian is closer to the actual world than the world where they are both French or that the world where they are both French is closer to the actual world than the world where they are both Italian. These two non-actual worlds seem to be *equally* close to the actual world. As a result, there does not seem to be any reason to treat one conditional as more acceptable than the other. So, more controversially, there is supposed to be no reason to suppose that the first conditional is to be regarded as true and the second as false or vice versa. However, let us look more closely at how this problem arises and why Stalnaker responds to the Bizet/Verdi case in the way that he does.

Stalnaker and Lewis independently proposed accounts of the logic of counterfactuals in the late 1960s and early 1970s. While these two theories are very similar formally, they were presented on the basis of somewhat different semantic ideas. Nevertheless, these semantic differences are largely superficial, with the exception of one major point of disagreement that in turn reflects a major difference regarding the formal principles characterizing the two different conditional logics they ended up endorsing. Let us begin by looking at the semantics for these two accounts of counterfactuals.

Stalnaker's semantics for counterfactuals was presented in terms of possible worlds and the concept of a selection function (Stalnaker, 1968). The selection function f takes a proposition and possible world pairs into a possible world. For Stalnaker, the truth conditions for counterfactuals are given as follows:

(C1) $A > B$ is true at world i, if and only if, B is true at $f(A, i)$.

Of course, f is governed by a number of well-known constraints.

Alternatively, Lewis's semantics for counterfactuals was presented in terms of a comparative similarity relation (Lewis, 1973). Where $S(i, j, k)$ means that j is more similar to i than k is to i, Lewis gives the truth conditions for counterfactuals as follows:

(C2) $A > B$ is true, if and only if, there is an A-world j such that B is true at j and in all A-worlds at least as similar to i as to j.

Stalnaker, however, showed that the choice of presenting semantics in terms of a selection function or in terms of a comparative similarity relation is really arbitrary (Stalnaker, 1981). Nevertheless, the two theories of counterfactuals that arise from these semantic bases and the constraints imposed on them are not strictly equivalent. The crucial point where the theories differ is that Stalnaker's theory assumes what Lewis called the limit and uniqueness assumptions. On the basis of these assumptions, Stalnaker endorses CEM. Lewis, however, disagrees and rejects the limit assumption and CEM. The details of the limit assumption are not important here, but acceptance of it and the uniqueness assumption is what gives rise to the problems associated with CEM noted above.[3] The uniqueness assumption can be stated as follows:

> (Uniqueness) For every world i and proposition A there is at most one A-world minimally different from i.

Accepting both of these assumptions amounts to the acceptance of CEM, but the uniqueness assumption effectively rules out ties in the similarity of worlds. If this principle is true, then there cannot be two worlds that are equally similar to a given possible world.

Stalnaker admits that this is an idealization that he has made with respect to the semantics of counterfactuals, specifically with respect to the selection function (Stalnaker, 1981: 89). Moreover, he defends this view on the basis of his own personal "unreflective linguistic intuition" (Stalnaker, 1981: 92) and argues essentially that treating both of the Bizet/Verdi counterfactuals as indeterminate in truth value better reflects these semantic intuitions than Lewis's view, where they both turn out to be false.

2 Coherence as a guide to counterfactual acceptance

Stalnaker and Lewis developed their semantic views of counterfactuals in terms of truth conditions, and both of their views were specifically framed in terms of possible worlds. However, we do not think the issue of the acceptability of CEM should turn on purely semantic considerations. Rather, what is needed is a clear account of the acceptability conditions for counterfactuals that explains the resistance to CEM and Bizet/Verdi type cases. Fortunately, there has been considerable discussion of this matter in the debate about the Ramsey test for conditional acceptance that is so-named because of Ramsey's brief footnote comment made in a paper in 1929.

In this vein, Carlos Alchourrón, Peter Gärdenfors, and David Makinson developed the AGM theory of belief revision in the 1980s and a number of related theories have arisen as a consequence (Alchourrón et al., 1985; Gärdenfors, 1988; Levi, 1996). The theory developed here will be specifically framed in terms of the version of this view presented in (Gärdenfors, 1988). These theories are fundamentally based on the concept of a belief state K, typically satisfying the following minimal conditions and where belief states are given a representation in some language L:

(BS) A set of sentences, K, is a belief state if and only if (i) K is consistent, and (ii) K is objectively closed under logical implication.

Given this basic form of epistemic representation, the AGM-type theories are intended to be a normative theory about how a given belief state satisfying BS is related to other belief states relative to: (1) the addition of a new belief b to K_i, or (2) the retraction of a belief b from K_i, where b ∈ K_i. Belief changes of the latter kind are *contractions*, but belief changes of the former kind must be further subdivided into those that require giving up some elements of K_i and those that do not. Additions of beliefs that do not require giving up previously held beliefs are *expansions*, and those that do are *revisions*.[4] Specifically, for our purposes here it is the concept of a revision that is of crucial importance to the issue of providing an account of rational commitment for conditionals. In any case, given AGM-style theories the dynamics of belief are simply the epistemically normative rules that govern rational cases of contraction, revision, and expansion of belief states.

The fundamental insight behind these theories is that belief changes that are contractions should be fundamentally conservative in nature. In other words, in belief changes one ought to make the minimal alterations necessary to incorporate new information and to maintain or restore logical consistency. This fundamental assumption is supposed to be justified in virtue of a principle of informational economy. This principle holds that information is intrinsically and practically valuable and so should be retained unless we are forced to do otherwise. So, while the details are not important here, the revision operations on belief states are restricted so as to obey such a principle of minimal mutilation.

What is important to the topic of this chapter is that on the basis of such theories of belief revision, the defenders of this approach to belief dynamics have also proposed that one could also give a theory of rational conditional commitment (Gärdenfors, 1981, 1982, 1988). The core concept of this theory is the Ramsey Test (Ramsey, 1929)[5]:

(RT) Accept a sentence of the form A > C in the state of belief K if and only if the minimal change of K needed to accept A also requires accepting C.[6]

Even in this quasi-formal form we can see what these theorists have in mind. The Ramsey Test requires that we modify our beliefs by accepting A into our standing system of beliefs and then see what the result is.[7] This view is typically framed in terms of a version of the epistemological coherence theory of justification and this seems natural given BS.[8] The idea is that one's beliefs are justified to the degree that they hang together or are mutually supportive. The idea then is that our belief system is justified in virtue of this feature of the system as a whole, and there are several extant versions of coherence theory that are plausible views of justification.[9] The most famous are of course those of Laurence BonJour and Keith Lehrer, but Paul Thagard's and Ted Poston's versions are also well regarded and more recent versions of coherentism (see BonJour, 1985; Lehrer, 1990; Thagard, 2000; Poston, 2014). In any case, we need not get bogged down in the debate about the particular details of coherentism here and we can simply adopt a basic, largely unanalyzed and broadly intuitive conception of coherence for the purposes of this chapter. This is also desirable because the results here are then not dependent on any particular version of coherence theory and so we shall simply accept that a belief state is coherent to the degree that its elements fit together and are mutually supportive. Once we accept this interpretation of RT and the notion of a belief state on which it is based, there is a natural way to extend RT to cases of comparative acceptance for conditionals in general and for Bizet/Verdi cases in particular.[10]

First, it is important to note that it is not at all clear that on RT either BV1 or BV2 is acceptable. This is because the minimal change of belief needed to incorporate the claim that Bizet and Verdi are compatriots does not obviously *require* accepting either that Bizet would have been Italian or that Verdi would have been French. But, both BV1 and BV2 seem to be acceptable conditionals nonetheless because accepting the shared antecedent permits one to accept either that Bizet would have been Italian or that Verdi would have been French. What is most important to recognize in the case of BV1 and BV2 is that they compete in an important sense. We then need to introduce the appropriate concept of a competitor as it applies to counterfactual conditionals. For the purpose of this chapter we can simply adopt the following concept of the competition of conditionals:

(COMP) A counterfactual conditional A > C competes with all other counterfactual conditionals that have A as an antecedent.

So, in the case of the Bizet/Verdi conditionals, we have a case of two competing conditionals and this should be no surprise. As we have already seen there is something important about the relationship between those two conditionals that ties them together intimately. Given COMP we can then replace RT with an appropriate concept of comparative acceptance given the coherentist interpretation of belief states as follows:

(CCA) Accept a sentence A > C in the state of belief K rather than A > B if and only if the minimal change of K needed to accept A, K', permits accepting C, the minimal change of K needed to accept A, K", also permits accepting B and the changes necessary to maintain the coherence of K' are less extensive than those necessary to maintain the coherence of K".

So defined, the principle of comparative conditional acceptance allows us to introduce a differential notion of conditional acceptance that is normative because it is based on the coherence theory of justification. Moreover, it allows us to explain Bizet/Verdi cases *without* having to introduce vagueness into the semantics for those conditionals.[11]

So why are our two conditionals so problematic and how does CCA make sense of the apparently problematic nature of them? Recall the Bizet/Verdi conditionals:

(BV1) If Bizet and Verdi had been compatriots, Bizet would have been Italian.
(BV2) If Bizet and Verdi had been compatriots, Verdi would have been French.

By COMP, BV1 and BV2 are competing counterfactual conditionals. Now if we apply CCA to our pair of sentences we should see that the revision of our state of belief K by the addition of the shared antecedent of BV1 and BV2 permits the acceptance of the claim that (I) Bizet would have been Italian and it also permits the acceptance of the claim that (F) Verdi would have been French.[12] This can be made more apparent by comparing the case of BV1 and BV2 with the cases where BV1 and BV2 are compared in terms of CCA with the following conditional:

(BV3) If Bizet and Verdi had been compatriots, Bizet would have been Dutch.

The changes necessary to accept BV3 are clearly more extensive than those needed to maintain consistency given the acceptance of BV1 or BV2. Moreover, given the relevant parts of our belief state and our intuitive understanding of coherence it is also reasonable to suppose that the revision of K by I, K', and the revision of K by F, K", are *equally extensive*. Both resultant belief states hang together or are mutually supportive to the same degree—or to a very similar degree—given what we know about Bizet, Verdi, and the world in general and the degree of change necessary to incorporate the antecedent and consequent of both is not noticeably different. It is just as coherent and requires the same sorts of changes of the same degree to suppose that, if the two men were compatriots, Bizet would be French as it is to suppose that, if the two men were compatriots, Verdi would be Italian. But the changes necessary to pursue either of these options in a coherent manner are clearly less extensive than the changes necessary to entertain the supposition that if the two men were compatriots, Bizet (or Verdi) would have been Dutch. Importantly, this means that while both BV1 and BV2 are acceptable there is no reason to accept BV1 over BV2 and no reason to accept BV2 over BV1 as per CCA. This then straightforwardly *explains* our inability to determine which is true and it explains this without any appeal to semantic vagueness. We do not need to take Stalnaker's radical semantic steps in order to deal with these sorts of cases. If the belief revision theory of counterfactual acceptance presented here is even broadly correct, then that the Bizet/Verdi cases are odd may well just be a reflection of a purely epistemic phenomenon and nothing deeper. This recognition in turn then shows that the Bizet/Verdi type cases do not decide the issue of CEM one way or the other. The *metaphysical/semantic* matter of whether there can be ties in terms of the similarities of worlds is not decided simply because we cannot *epistemically* distinguish conditionals in Bizet/Verdi type cases, and in deference to the principle of minimal mutilation we ought to resist the move to introduce vagueness into the semantics of conditionals *pace* Stalnaker.

The foregoing theoretical considerations can be supplemented by empirical evidence concerning our shared practices of asserting and evaluating counterfactual conditionals. To this end, we undertook three empirical studies of folk judgments about counterfactuals and CEM, which we describe in the following sections. Collecting both quantitative and qualitative data, we found that nonexperts' intuitive judgments about Bizet/Verdi-style counterfactuals and CEM accord better with the belief revision view than with the Lewisian or Stalnakerian views. We do not believe that data about folk judgments can be decisive in debates like the present one or that they should trump more theoretical considerations. Nevertheless, we believe the data we report below represent an important kind of consideration for theorists to take into account—particularly since the target of investigation is the ordinary meaning of an everyday use of language. Together, we maintain that the theoretical and empirical considerations we adduce provide reason for preferring the belief revision view over the Lewisian and Stalnakerian views.

3 Study 1

In Study 1, we presented 150 participants (34 percent female, average age = 34, predominantly Caucasian, 97 percent native English speakers, all located in the United States) who were recruited via Amazon's Mechanical Turk (www.mturk.com) with the following vignette[13]:

> *Neighbors.* Joe is a Minnesotan who has always lived in Minneapolis, Minnesota, and Jane is a New Yorker who has always lived in Buffalo, New York. Since they have always lived in different states, they have never lived in the same neighborhood. They have never met or talked to one another. They don't even know about the existence of the other.

Participants were randomly separated into two conditions and were asked to think about two different kinds of Bizet/Verdi-style counterfactuals concerning Joe and Jane. Those in the "were/would be" condition received the following instructions:

> Now think about what would be true if Joe and Jane were neighbors. Please read each of the following statements and select the best description of each statement that follows.
>
> (1.1) If Joe and Jane were neighbors, then Joe would be a New Yorker.
> (1.2) If Joe and Jane were neighbors, then Jane would be a Minnesotan.
> (1.3) If Joe and Jane were neighbors, then Jane would be a Texan.

The order in which these statements were presented was counterbalanced.

Statements (1.1) and (1.2) are Bizet/Verdi-style counterfactuals, since there does not seem to be any reason to suppose that the closest worlds where Joe and Jane are neighbors are worlds where Joe is a New Yorker rather than worlds where Jane is a Minnesotan (or vice versa). Statement (1.3) was included for the sake of comparison, since it is clear that there are many worlds closer to the actual one in which Joe and Jane are not Texans.

After each statement, participants were given the following answer choices (always in the following order):

> __ I think this statement is true.
> __ I think this statement is false.

__ I think this statement is both true and false at the same time.
__ I think this statement is neither true nor false.
__ I think this statement is either true or false. I just don't know which one it is.

After each set of answer choices, participants were prompted to explain why they chose the answer they did. All of the above questions (in addition to demographic questions) were presented on a single webpage.

Participants in the "had been/would have been" condition received the following instructions:

> Now think about what would be true if Joe and Jane had been neighbors. Please read each of the following statements and select the best description of each statement that follows.
>
> > (1.4) If Joe and Jane had been neighbors, then Joe would have been a New Yorker.
> > (1.5) If Joe and Jane had been neighbors, then Jane would have been a Minnesotan.
> > (1.6) If Joe and Jane had been neighbors, then Jane would have been a Texan.

Statement order was counterbalanced, and the answer choices were the same as above. Statements (1.4) and (1.5), like Statements (1.1) and (1.2), are Bizet/Verdi-style counterfactuals. Standard accounts of counterfactuals do not treat "were/would be" counterfactuals differently than "had been/would have been" counterfactuals. However, because we did not want to presume in advance that folk judgments about the two kinds of counterfactuals would be the same, we wanted to make both kinds of counterfactuals available to participants.

We did not expect participants to select "I think this statement is true" very often for any of the Bizet/Verdi-style counterfactuals, since the vignette provides no reason for preferring any one of them to its counterpart. We also did not expect participants to endorse "I think this statement is both true and false at the same time," since we did not expect dialetheism to be common among the folk. If participants were to make judgments in line with the Lewisian perspective, they should select "I think this statement is false" for each of the Bizet/Verdi-style counterfactuals. The response "I think this statement is neither true nor false" operationalized the Stalnakerian perspective in the present study. The belief revision option was represented by "I think this statement is either true or false. I just don't know which one it is."

Table 11.1 Distributions of participant response choices for Statements 1.1 through 1.6 in Study 1, with the most commonly selected responses highlighted

	True (%)	False (%)	Both T & F (%)	Neither T nor F (%)	Either T or F (%)
Statement 1.1	8.3	23.6	26.4	13.9	27.8
Statement 1.2	14.7	24.0	21.3	16.0	24.0
Statement 1.3	2.7	76.0	2.7	5.3	13.3
Statement 1.4	12.2	14.9	28.4	17.6	27.0
Statement 1.5	13.3	14.7	28.0	14.7	29.3
Statement 1.6	1.4	81.1	1.4	9.5	6.8

We hypothesized that participants would choose the belief revision option more often than the others for Bizet/Verdi-style counterfactuals. Participant responses are summarized in Table 11.1, with the most commonly selected responses for each question highlighted.[14]

Chi-squared goodness-of-fit tests were performed on participants' responses to each statement in order to see if each distribution of answer choices differed significantly from chance. The tests were significant for Statements (1.1), (1.3), and (1.6) and approached significance for (1.5).[15] In order to see whether participants treated the "were/would be" counterfactuals and the "had been/would have been" counterfactuals differently, we ran three chi-square tests of independence on participants' answers to Statements (1.1) and (1.4), (1.2) and (1.5), and (1.3) and (1.6). In each case, we observed no statistically significant difference in the distribution of participants' answer choices.[16] Thus, in what follows, we do not draw any important distinctions between "were/would be" counterfactuals and their "had been/would have been" counterparts.

Participants overwhelmingly responded to Statements (1.3) and (1.6) by giving the intuitively correct verdict "False." Because the three theoretical perspectives under consideration do not differ in regard to the correct response to these counterfactuals, we set these data aside for now. Collapsing data across the "were/would be" and the "had been/would have been" conditions, we find that in response to Statements (1.1), (1.2), (1.4), and (1.5), participants gave the Lewisian answer ("False") 19.3 percent (±4.5 percent) of the time, the Stalnakerian answer ("Neither") 15.5 percent (±4.1 percent) of the time, and the belief revision response 27.0 percent (±5.1 percent) of the time.[17]

Participants selected "I think this statement is true" in response to the Bizet/Verdi-style counterfactuals 12.2 percent of the time. The Lewis, Stalnaker, and belief revision perspectives all agree that this is an incorrect response to give. Interestingly, of those participants who provided explanations for why they chose this answer, it was most often the case that participants explain they are not actually fully committed to these statements being true. In 64 percent of these explanations, participants said that they were thinking there must be in reality some fact that breaks the tie between the closest possible worlds. The following explanations are typical of this kind of response:

> Since Jane is from NY, any neighbor of hers theoretically must be a New Yorker. So this would be true. [Statement 1.4]
> Joe has always lived in Minnesota, so for a neighbor to be actually a neighbor, that person would be a Minnesotan. This includes Jane. [Statement 1.5]

Fifteen percent of the explanations for why "I think this statement is true" was selected in response to the Bizet/Verdi-style counterfactuals employed modal terms to suggest there might not ultimately be a tie between the closest relevant possible worlds. For example:

> Jane is from New York so it is possible for them to live there. [Statement 1.1]
> Joe is from Minnesota so it is possible for them to live there. [Statement 1.2]

In other words, these explanations reveal that these participants meant "I think it is possible that this statement is true." These explanations suggest that if more were known about the situations in question, there would likely not be a tie in the closest possible worlds and it would be clear whether they were true or false.[18]

Several of the 19.2 percent of participants who selected the Lewisian answer ("False") in response to the Bizet/Verdi-style counterfactuals indicated in their explanations that they selected this answer because they treated "New Yorker" or "Minnesotan" as terms of personal identity rather than as terms that denoted where someone lived:

> Even if Joe lived in New York, he would be a Minnesotan since he was from there. [Statement 1.1]
> No matter what state [Jane] moves to now, she would still be considered a New Yorker. [Statement 1.5]

Fourteen percent of participants' explanations for why they selected this answer choice unambiguously treated "New Yorker" and "Minnesotan" in this fashion. These interpretations of the terms "New Yorker" and "Minnesotan" significantly diverge from how we intended them to be understood. In subsequent studies, we made sure to use terms that could not be misinterpreted in this way. When we subtract those responses that were clearly based upon a misinterpretation of "New Yorker" or "Minnesotan," we see that participants expressed agreement with the Lewisian perspective at most 16 percent of the time.

Forty-six percent of the explanations for why "I think this statement is false" was chosen in response to the Bizet/Verdi-style counterfactuals appealed in some way to the fact that the information provided in the Neighbor vignette does not provide any reason for thinking these statements are true. Since it is specified that Joe has always lived in Minnesota, these participants did not see how the consequents of (1.1) and (1.4)—viz., "Joe would be a New Yorker" and "Joe would have been a New Yorker"—could be true. An additional 18 percent of explanations were an ambiguous combination of the first two kinds of explanation—meaning that the true percentage of participants who actually agree with Lewis is probably lower than the estimate given above.[19]

Participants selected the answer "I think this statement is both true and false at the same time" 26.0 percent of the time in response to the Bizet/Verdi-style counterfactuals. An analysis of the explanations of their reasons for selecting this answer reveals that 0 percent of the participants who chose this response actually think the counterfactuals are literally both true and false at the same time. Instead, 87 percent of these participants clearly and unambiguously explained that they chose this answer because of the way it depicted a tie between the considerations in favor of saying the counterfactual in question was true and saying it was false[20]:

> It's possible that Joe could be a New Yorker this way, but it's just as possible they could have both been neighbors in Minnesota instead. [Statement 1.1]
>
> This statement could be true and false at the same time because it depends on if they were neighbors in New York or Minnesota. [Statement 1.1]
>
> I have no info about where Joe is living at the time if they were neighbors. [Statement 1.1]
>
> It could go either way; Joe lives in Minnesota, so Joe could also be a Minnesotan if Jane was his neighbor. Not enough info to know who must have moved. [Statement 1.4]
>
> It really depends on who moved where to figure who is what. [Statement 1.5]

Almost every participant who selected "I think this statement is both true and false at the same time" used modal terms to explain that they thought of this answer as expressing an equal possibility between truth and falsity—not that they thought the statement in question was actually both true and false at the same time. Thus, it seems that all of the participants who selected this third answer endorse either the Stalnakerian view that the facts of the case are such that it is indeterminate whether the Bizet/Verdi-style counterfactuals are true or false or the belief revision view that there is a fact of the matter about which we are ignorant. Participants' answers strongly suggest they disagree with the Lewisian view that such counterfactuals are false.

Participants selected "I think this statement is neither true nor false" 15.5 percent of the time in response to the Bizet/Verdi-style counterfactuals. Almost every explanation participants provided for their answer choices focused on the fact that there was no way for them to know whether Joe would be a New Yorker or Jane would be a Minnesotan, if Joe and Jane were neighbors:

> I don't have enough information to make a determination on whether this statement is true or false. [Statement 1.1]
> Could have lived in either state, neighbors where isn't specified. [Statement 1.1]
> This hypothetical doesn't suggest where the two might live if they were neighbors. [Statement 1.4]

The most natural way to read all of the explanations participants provided of why they selected this answer choice is to see them as making epistemological points. It is not that they literally think the counterfactuals lack truth values and are actually neither true nor false. Rather, it is that they do not possess any reason for thinking the statements have one truth value rather than another. However, the idea of a statement being neither true nor false might be a rather difficult thing for an M-Turk worker to express even if they wanted to endorse such a view. Therefore, in subsequent studies described below, we attempted to make it easier for participants to express such an idea.

Participants selected "I think this statement is either true or false. I just don't know which one it is" 27.0 percent of the time in response to the Bizet/Verdi-style counterfactuals. The explanations they gave were entirely uniform and seemed to express a commitment to the idea that whatever facts made Joe and Jane neighbors would also make both of them New Yorkers or both of them Minnesotans. The participants simply noted that they were not privy to any information about those facts:

> There's a lack of information. If they had been neighbors either Joe has been a New Yorker or Jane has been a Minnesotan. [Statement 1.1]
> It would have to depend on how they ended up being neighbors. [Statement 1.2]

There are two scenarios where they can be neighbors. They can be neighbors in either Minnesota, thus making Jane a Minnesotan or they can be neighbors in New York. Therefore, the statement is true when they are neighbors in Minnesota, but not when they are in New York. [Statement 1.5]

These responses seem to express something like the belief revision view, although as we noted above it could be difficult for participants to articulate a view like Stalnaker's even if they wanted to.

The most direct and natural interpretation of the quantitative and qualitative data from Study 1 is that participants' judgments on the whole express something closer to the belief revision account of counterfactuals than either of the other views we considered above. Participants' judgments thus provide us with no reason to reject CEM or to endorse supervaluationism for Bizet/Verdi-style counterfactuals.

4 Study 2

In our second study of nonexperts' judgments about Bizet/Verdi-style counterfactuals, we constructed a vignette that had the same structural or logical features as Neighbor but that avoided using terms like "New Yorker" or "Minnesotan" that could be interpreted as identity terms. For Study 2, we recruited seventy-five workers (51 percent female, average age = 38, predominantly Caucasian, 100 percent native English speakers, all located in the United States) from Amazon's Mechanical Turk[21] to read and respond to the following vignette:

> *Same Religion.* Jordan is a devout Christian who was raised in a Christian household and has practiced Christianity her entire adult life. Tenzin is a devout Buddhist who was raised in a Buddhist household and has practiced Buddhism his entire adult life. Jordan has never considered practicing Buddhism. Tenzin has never considered practicing Christianity.

Participants were given the following instructions:

> Now think about what would be true if Jordan and Tenzin practiced the same religion. Please read each of the following statements and select the best description of each statement that follows.
>
> (2.1) If Jordan and Tenzin practiced the same religion, then Jordan would be Buddhist.

(2.2) If Jordan and Tenzin practiced the same religion, then Tenzin would be Christian.

(2.3) If Jordan and Tenzin practiced the same religion, then Jordan would be Muslim.

Procedures and answer choices were the same as in Study 1. Participant responses are summarized in Table 11.2.

Chi-squared goodness-of-fit tests were performed on participants' responses to each question in order to see if each distribution of answer choices differed significantly from chance. Each test was significant.[22] In other words, the distributions of participant answer choices differed significantly from a (flat) distribution in which each answer choice was chosen equally often. Participants selected the Lewisian response 12.2 percent (±5.3 percent) of the time, the Stalnakerian view 15.0 percent (±5.8 percent) of the time, and the belief revision view 41.5 percent (±8.0 percent) of the time.

For both of the Bizet/Verdi-style counterfactuals, participants strongly preferred the fifth answer choice, viz., "I think this statement is either true or false. I just don't know which one it is." We hypothesize that this difference in the distribution of responses observed in Studies 1 and 2 is due to the fact that in Study 2 it seemed clearer to participants that one of the relevant states of affairs at issue (e.g., practicing Buddhism or being a New Yorker) precluded the other (e.g., practicing Christianity or living in Minnesota).

Participants' explanations for why they chose the answers they did were virtually identical to those observed in Study 1.[23] The few who selected "I think this statement is true" often used modal terms to explain that they thought the statement could be true. Those who selected "I think this statement is false" cited the lack of information in the vignette for thinking either of the statements are true. Participants who chose "I think this statement is both true and false at the same time" said they did so because there was no reason to give one of the statements more credence than the other. And so on.

Table 11.2 Distributions of participant response choices for Statements 2.1 through 2.3 in Study 2, with the most commonly selected response for each question highlighted

	True (%)	False (%)	Both T & F (%)	Neither T nor F (%)	Either T or F (%)
Statement 2.1	10	12	21	15	42
Statement 2.2	12	12	20	15	41
Statement 2.3	3	57	8	9	23

Again, the most straightforward interpretation of these quantitative and qualitative data is that participants' judgments about Bizet/Verdi-style counterfactuals accord with the belief revision account of counterfactuals better than the Lewisian or Stalnakerian accounts and thus provide no reason to reject CEM or to endorse supervaluationism.

5 Study 3

Studies 1 and 2 have certain features that might make them less than ideal probes of participants' intuitive judgments about Bizet/Verdi-style counterfactuals. One such feature is that the five answer choices participants were asked to choose from were not all uniform. Each of the first four options ("I think this statement is true," "I think this statement is false," "I think this statement is both true and false at the same time," and "I think this statement is neither true nor false") consists of a single sentence, whereas the fifth option ("I think this statement is either true or false. I just don't know which one it is") consists of two sentences. Furthermore, the second sentence contains the phrase "I just don't know," which might attract more uncertain participants to select this option than actually agreed with the semantic thesis stated in the initial sentence.

A second feature shared by Studies 1 and 2 is that, while the five answer choices they employ embodied different perspectives on Bizet/Verdi-style counterfactuals (and, by implication, CEM), they did not articulate any rationale one might have for selecting one of these options. Of course, participants were asked to explain their rationales. But we thought that providing them with different rationales to choose from might shed additional light on underlying factors driving their judgments.

A third feature is that the Neighbor and Same Religion vignettes feature pairs of counterfactual conditionals whose consequents are contrary to one another and do not exhaust the full range of possibilities. CEM, however, is stated in terms of a pair of conditionals in which the consequent of one is the negation of the consequent of the other. "Joe would be a New Yorker" is not the negation of "Jane would be a Minnesotan" (or vice versa). And one can reside somewhere besides New York or Minnesota. The same is true of being Buddhist and being Christian. Of course, "Bizet would have been Italian" is not the negation of "Verdi would have been French" either. And being Italian and being French do not exhaust all the nationality possibilities. So, we do not think it was out of bounds for us to use the conditionals we did. Furthermore, Lewis would say that each of the Bizet/Verdi-style counterfactuals we have considered above—viz., (BV1), (BV2), (1.1), (1.2), (1.4), (1.5), (2.1), and (2.2)—is false and hence that CEM must be false as well. Thus, the Bizet/Verdi-style counterfactuals we employed have direct implications for the merits of CEM. Nonetheless, because the law of excluded middle is stated in terms of consequents that exhaust the full range of possibilities in which the antecedent is true, we wanted to see if we would obtain the same pattern of intuitive judgments with pairs of conditionals whose form more closely matched that of CEM.

Therefore, participants in Study 3 were asked to read the following vignette:

Hemisphere. Maddi lives in the northern hemisphere. She has lived there her entire life and has never even visited the southern hemisphere. Aleah lives in the southern hemisphere. She has lived there her entire life and has never even visited the northern hemisphere. Maddi has never considered moving to the southern hemisphere, and Aleah has never considered moving to the northern hemisphere.

Participants were then given the following instructions:

Now think about what would be true if Maddi and Aleah were to live in the same hemisphere. Which (if any) of the following statements would be true?

(3.1) If Maddi and Aleah were to live in the same hemisphere, then they would both live in the northern hemisphere.
(3.2) If Maddi and Aleah were to live in the same hemisphere, then they would both live in the southern hemisphere.

There are three main schools of thought about how to view statements like these:

View 1 says that one of these statements must be true and one of them must be false because there is always something that breaks apparent ties between statements like these and makes one of them true and the other one false. This remains the case even when we can't tell what the tie-breaker is or which statement is the true one.

View 2 says that it is possible there is nothing that breaks the apparent tie between these statements and that if there is a genuine tie, both statements are false. Truth requires that the facts favor one of the statements over the other. But if the facts don't favor one statement over the other, they cannot be true. The only other option is for both of them to be false.

View 3 says that the facts of the case described above are not settled enough for the statements to be either true or false. Both truth and falsity require the facts to be settled in favor of one statement or the other. But when the facts are unsettled, there is nothing that makes the statements either true or false.

The order of presentation of Views 1 through 3 was counterbalanced across participants. View 1 corresponds to the belief revision account of counterfactuals and CEM, View 2 is Lewisian, and View 3 is Stalnakerian. In contrast to the Bizet/Verdi-style counterfactuals used in Studies 1 and 2, (3.1) and (3.2) feature consequents that exhaust the full range of possibilities. Instead of simply presenting participants with options like "I think this statement is true," "I think this statement is false," etc., with no accompanying rationale, the three views described above pair judgments about the truth values of (3.1) and (3.2) with explanations for those verdicts.

Participants were then asked the following seven comprehension questions. Correct answers are marked in bold.

(4.1) Maddi lives:
  __ **in the northern hemisphere**
  __ in the southern hemisphere

(4.2) Aleah lives:
  __ in the northern hemisphere
  __ **in the southern hemisphere**

(4.3) According to View 1, one of the statements (1) and (2) must be true.
  True
  False

(4.4) According to View 2, one of the statements (1) and (2) must be true.
  True
  False

(4.5) According to View 3, one of the statements (1) and (2) must be true.
  True
  False

(4.6) According to View 2, statements (1) and (2) are both false.
  True
  False

(4.7) According to View 3, statements (1) and (2) are both false.
  True
  False

Finally, participants were asked the following key questions:

(4.8) The view I find most plausible is:
　　View 1
　　View 2
　　View 3
　　I don't know

(4.9) Please explain your answer

All of the above questions (in addition to demographic questions) were presented on a single webpage.

Three hundred participants (46 percent female, average age = 38, predominantly Caucasian, 99 percent native English speakers) were recruited via Amazon's Mechanical Turk to take part in Study 3.[24] Their responses are summarized in Table 11.3.

A chi-square test of independence revealed a statistically significant difference between the distribution of answers to (4.8) of those participants who answered all of the comprehension questions correctly and the distribution of those who did not.[25] The primary driver of this difference was that fewer participants who answered all comprehension questions correctly selected "I don't know" than those who did not. As in Studies 1 and 2, participants exhibited a marked preference against the Lewisian view. Three quarters of participants selected either the belief revision or the Stalnakerian view, with a modest edge going to the Stalnakerian view.

When we examine participants' explanations for why they selected the answers they did, we find striking discontinuities in how participants understood the relevant views. The 102 participants who selected View 1 (the belief revision view) as the most plausible option offered basically identical

Table 11.3 Distributions of participant answers to Statements 4.8 in Study 3, organized by whether participants answered all seven comprehension questions correctly. The most commonly selected responses in each column are highlighted

	All correct ($n = 151$) (%)	Not all correct ($n = 149$) (%)	Total ($n = 300$) (%)
Belief Revision	33.8 (±7.5)	34.2 (±7.6)	34.0 (±5.4)
Lewis	16.6 (±5.9)	16.8 (±6.0)	16.7 (±4.2)
Stalnaker	47.0 (±8.0)	33.6 (±7.6)	40.3 (±5.6)
Don't Know	2.6 (±2.5)	15.4 (±5.8)	9.0 (±3.2)

explanations for their choice—explanations that clearly accord with the intended understanding of View 1:

> It just makes the most sense for one to be true and one to be false. I feel that if they live in the same hemisphere, one of the two statements has to be true, though I cannot determine which one. There is no way that both of the statements is false because there are only two hemispheres. If they live in the same one, statement 1 or 2 has to be true because there is no other option.
>
> One of them has to be true. If they are living in the same hemisphere it has to be one or the other.
>
> The statement says that they both live in the same place, so one must be true, although there is no way to tell which one is true.
>
> Even though I don't know which hemisphere they are together in, they have to be in one or the other.

However, matters were markedly different with the 121 participants who selected View 3 (the Stalnakerian view). Only 2.5 percent of them unambiguously expressed a commitment to the idea that statements (3.1) and (3.2) lack truth values. Almost every explanation unambiguously articulated the idea that the choice between (3.1) and (3.2) was epistemically underdetermined:

> I don't think we have all the information we need to correctly assess the situation.
>
> There isn't enough information to confirm or deny either statement.
>
> I don't know enough of the extenuating circumstances surrounding the girls to know what would happen, so they can be neither true nor false.
>
> Because there are not of facts to prove one way or another.
>
> There is not enough evidence to suggest either way and until the facts are settled, there is no way to predict.
>
> I feel like in order for these statements to be true or false, more information would need to be given to base that answer off of. I don't think either is necessarily fully true or false.
>
> We don't know that either of the two people mentioned would move to the location of the other. So, not enough information to determine.
>
> Since we don't know which person would move, we can't say for sure which hemisphere they live in.
>
> There isn't enough information to decide on the answer true or false.
>
> Because there isn't enough info to know which is true and which is false.
>
> You can't establish truth or falsehood without the facts on which to base a conclusion.
>
> I don't think either statement is necessarily true or false because there are no real facts that settle the argument.

> There is nothing stated that makes me believe and come to the conclusion on which is true and which is false
>
> It can't be that both are false because you HAVE to live in a hemisphere. I there are factors that would settle the question.

These explanations employ epistemic terms like "don't know," "information," "prove," "assess," "confirm," "deny," "determine," "say for sure," "decide," "establish truth or falsehood," "conclusion," "argument," "believe," and "settle the question." View 3 (the Stalnakerian view) is supposed to be a view on which matters are unsettled in a semantic sense rather than in an epistemic sense—one where the facts of the case fail to make (3.1) and (3.2) true or false. On this view, not even god could know that (3.1) or (3.2) is true because there is no truth to be known. Yet participants who chose View 3 almost unanimously offered explanations in which the unsettled nature of the case was epistemic—one where there are facts but these facts remain unknown.

Thus, despite what the quantitative data from Question (4.8) initially suggest, the qualitative data from Question (4.9) significantly undermine the conclusion that folk judgments about counterfactuals and CEM accord more with the Stalnakerian view than the belief revision view.

6 Conclusion

In this chapter we considered the relevant details of three philosophical perspectives on the nature of counterfactual conditionals, particularly as they apply to Bizet/Verdi cases. Importantly, this includes the perspectives of both Stalnaker and Lewis. We examined in some detail their views on the question of whether and how the principle of conditional excluded middle figures in the logic and semantics of counterfactuals. Following this expository project, we articulated and defended a third view based on belief revision theories in the tradition of the Ramsey Test. Unlike Lewis's view, the belief revision perspective does not reject conditional excluded middle and does not treat such competing conditionals as false. Unlike Stalnaker's view it does not embrace supervaluationism and treat such competing conditionals as indeterminate in truth value. On this basis we adduced what we think are compelling theoretical reasons in favor of the belief revision view. However, we also reported the results of three empirical studies of nonexperts' judgments about counterfactuals and conditional excluded middle that provide additional empirical support for the belief revision view. We conclude then that it is likely the seeming difficulties associated with Bizet/Verdi cases are merely epistemic and thus unlikely that they are driven by deeper semantic considerations.

Notes

1 The exposition in the opening sections of this chapter closely follows Shaffer (2016).
2 Lewis also effectively appeals to the supervaluational theory, but he disagreed with Stalnaker about CEM and its basis in the possible world's framework.
3 See (Cross, 2009) for discussion of the relationship between the limit assumption, the uniqueness assumption, and the principle of counterfactual consistency.
4 In point of fact the AGM theory really only holds that there are two dynamical operations on belief states, because revision is defined in terms of expansion and contraction.
5 See (Shaffer, 2011, 2013) for some discussions of problems for naïve formulations of the Ramsey Test.
6 For a relatively recent discussion of RT and related views, see (Levi, 2004).
7 (Sanford, 2003) contains the objection that in many cases where the antecedent of such a conditional is a radical departure from what we believe to be the case, we cannot in fact employ the Ramsey Test because we do not know what would be the case if we believed such an antecedent. So, he claims that many conditionals are simply void, rather than true or false. It is worth pointing out here that Sanford's criticism is weak at best. It simply does not follow that because we cannot always clearly determine what would be the case if we were to believe some claim, a conditional with such an antecedent has no truth value. See (Williamson, 2007: chapters 5 and 6) for discussion of one suggestion for how such knowledge might be obtained.
8 See (Gärdenfors, 1992) for the most thorough defense of the AGM theory in terms of coherentism. See (Shaffer, 2002) for some worries about this view.
9 There is of course some controversy about such views, especially those that are framed in terms of probabilistic notions of coherence. See (Bovens and Hartmann, 2003; Olsson, 2005) for discussion of this matter.
10 We acknowledge here both that our invocation of AGM-type theories as a basis for an epistemic theory of conditional acceptance is not wholly worked out here and that it is potentially problematic. Gärdenfors himself pointed out in (Gärdenfors, 1986) and in (Gärdenfors, 1988: chapter six) that the AGM models for conditional acceptance are trivial. But, we think that the general approach is sound, as does Gärdenfors himself (1988: 166).

11 The reliability of semantic intuitions has recently been questioned in (Machery et al., 2004).
12 This can be seen also in that both BV1 and BV2 satisfy RT.
13 Workers were required to have an approval rating of at least 98 percent on at least 5000 M-Turk tasks. No worker was permitted to participate in more than one condition or more than one study. Workers were paid $.35 for their participants in Study 1.
14 Complete data sets from each of our studies can be found here: https://osf.io/a5jty/.
15 (1.1): $\chi^2(4, N = 72) = 10.36, p = .035$. (1.2): $\chi^2(4, N = 75) = 2.93, p = .57$. (1.3): $\chi^2(4, N = 75) = 149.87, p < .001$. (1.4): $\chi^2(4, N = 74) = 7.89, p = .096$. (1.5): $\chi^2(4, N = 75) = 9.47, p = .050$. (1.6): $\chi^2(4, N = 74) = 174.38, p < .001$.
16 (1.1) & (1.4): $\chi^2(4, N = 146) = 2.35, p = .672$. (1.2) & (1.5): $\chi^2(4, N = 150) = 2.86, p = .582$. (1.3) & (1.6): $\chi^2(4, N = 149) = 3.22, p = .521$.
17 The percentages in parentheses represent 95 percent confidence intervals.
18 The remaining 21 percent of explanations did not provide a clear rationale for why "I think this statement is true" was chosen. These explanations include "They would be in the same state" [Statement 1.1] and "Because Joe and Jane were neighbors" [Statement 1.5].
19 The remaining 5 percent of explanations were insufficiently clear to permit confident categorizations.
20 Four percent of participants interpreted "New Yorker" and "Minnesotan" as identity terms. The remaining 9 percent of responses were not sufficiently clear to be able to place them in any category.
21 Workers were required to have an approval rating of at least 98 percent on at least 5000 M-Turk tasks. Workers were paid $.35 for their participants in Study 2.
22 (2.1): $\chi^2(4, N = 73) = 25.43, p < .001$. (2.2): $\chi^2(4, N = 74) = 21.14, p < .001$. (2.3): $\chi^2(4, N = 74) = 70.73, p < .001$.
23 The only difference was that no participant in Study 2 interpreted any of the key terms as denoting personal identity.
24 Workers were required to have an approval rating of at least 98 percent on at least 5000 M-Turk tasks. Workers were paid $.45 for their participants in Study 3.
25 $\chi^2(3, N = 300) = 17.00, p < .001$, Cramér's V $= .24$.

References

Alchourrón, C., Gärdenfors, P. and Makinson, D. (1985). On the logic of theory change: Partial meet functions for contraction and revision. *Journal of Symbolic Logic*, 50:510–30.

Bonevac, D. (2003). *Deduction*, second edition. Oxford: Blackwell.

BonJour, L. (1985). *The Structure of Empirical Knowledge*. Cambridge, MA: Harvard University Press.
Bovens, L. and Hartmann, S. (2003). *Bayesian Epistemology*. Oxford: Oxford University Press.
Cross, C. (2009). Conditional excluded middle. *Erkenntnis*, 70:173–88.
Gärdenfors, P. (1981). An epistemic approach to conditionals. *American Philosophical Quarterly*, 18:203–11.
Gärdenfors, P. (1982). Imaging and conditionalization. *The Journal of Philosophy*, 79:747–60.
Gärdenfors, P. (1986). Belief revisions and the Ramsey test for conditionals. *The Philosophical Review*, 95:81–93.
Gärdenfors, P. (1988). *Knowledge in Flux*. Cambridge, MA: MIT Press.
Gärdenfors, P. (1992). The dynamics of belief systems: Foundations versus coherence theories. In C. Bicchieri and M. Dalla Chiara (Eds.), *Knowledge, Belief and Strategic Interaction*, 377–96. Cambridge: Cambridge University Press.
Lehrer, K. (1990). *Theory of Knowledge*. Boulder: Westview Press.
Levi, I. (1996). *For the Sake of the Argument: Ramsey Test Conditionals, Inductive Inference, and Nonmonotonic Reasoning*. Cambridge: Cambridge University Press.
Levi, I. (2004). *Mild Contraction*. Oxford: Oxford University Press.
Lewis, D. (1973). *Counterfactuals*. Cambridge, MA: Harvard University Press.
Machery, E., Mallon, R., Nichols, S. and Stich, S. (2004). Semantics, cross-cultural style. *Cognition*, 92:B1–12.
Olsson, E. (2005). *Against Coherence*. Oxford: Oxford University Press.
Poston, T. (2014). *Reason and Explanation: A Defense of Explanatory Coherentism*. Basingstoke: Palgrave Macmillan.
Quine, W. V. (1950). *Methods of Logic*. New York, NY: Holt, Reinhart and Winston.
Ramsey, F. P. (1929/1990). Laws and causality. Reprinted in D. H. Mellor (Ed.), *F.P. Ramsey: Philosophical Papers*, 140–63. Cambridge: Cambridge University Press.
Sanford, D. (2003). *If P, then Q*, second edition. New York, NY: Routledge.
Shaffer, M. (2002). Coherence, justification, and the AGM theory of belief revision. In Y. Bouchard (Ed.), *Perspectives on Coherentism*, 139–60. Ontario: Aylmer-Éditions du Scribe.
Shaffer, M. (2011). Three problematic theories of conditional acceptance. *Logos & Episteme*, 1:117–25.
Shaffer, M. (2013). Doxastic voluntarism, epistemic deontology and belief-contravening commitments. *American Philosophical Quarterly*, 50:73–82.
Shaffer, M. (2016). What if Bizet and Verdi had been compatriots? *Logos & Episteme*, 7:55–63.
Stalnaker, R. (1968). A theory of conditionals. In N. Rescher, (Ed.), *Studies in Logical Theory*, 98–112. Oxford: Blackwell.

Stalnaker, R. (1981). A defense of conditional excluded middle. In W. Harper, R. Stalnaker and G. Pearce, (Eds.), *Ifs*, 87–104. Dordrecht, The Netherlands: D. Reidel.

Thagard, P. (2000). *Coherence in Thought and Action*. Cambridge, MA: MIT Press.

Van Fraassen, B. (1966). Singular terms, truth-value gaps and free logic. *Journal of Philosophy*, 63:481–95.

Williamson, T. (2007). *The Philosophy of Philosophy*. Oxford: Blackwell.

Index

Aberdein, Andrew 1–13, 30, 66–8, 82–5, 125
Adams, Ernest 237, 243
aequitas hermeneutica 25
aesthetics 3, 7, 63–9, 71, 73–4, 77–86, 88–9
Alchourrón, Carlos 9, 255
Alexander, Joshua 18
ANOVA analysis 156
approximate number system (ANS) 43–4, 47–9, 55
a priori 5, 169, 196, 202, 204, 205, 207, 211–21
argumentation theory 2, 5, 87
Aristotelian realism 54

Barnard, Robert 1, 8, 145–71
Beebe, James R. 9, 251–76
belief, degrees of 225, 227, 236, 238, 243
biconditional (equivalence schema; T-schema) 145–67
Bizet/Verdi conditionals 9, 251–4, 256–73
Brown, James Robert 8, 96, 98–101, 112, 117, 124–5
Buhrmester, Michael 169

calculus 126, 128–35, 137, 139
Carnap, Rudolf 31, 168
Chrysippus 5
CIELAB space 188
CIELUV space 188–9, 198
Clarke-Doane, Justin 42–3, 52–3
cluster concept 23, 25
coherence 236–7, 240, 254, 256–7, 273
color space 179, 184, 188–91, 195–6
common sense 25, 146–9, 202–3
conceptual spaces 9, 178–9, 184–5, 187–8, 191, 195, 197
conditional, indicative 225–6, 229–30, 232–5, 243

conditional, material 226–8, 230–1, 237, 239, 242, 244
conditional, probability 229, 237–9, 242
conditional excluded middle, principle of 9, 251–2, 254, 258, 265, 267–9, 272–3
conditional probability 228–43
convention-T 145–9, 160, 165
counterexamples 25, 28, 74–7, 87, 100, 106, 110–18
counterfactual 234–5, 241–4, 251–4, 256–69, 272
Cruz, Nicole 9, 225–49

Darwin, Charles 40–1, 56, 151–4, 158–9, 168
data mining 20, 201, 206, 219–20
data science 201, 219–21
Davidson, William Leslie 169
debunking arguments 7, 39, 42, 52
Decock, Lieven 176, 178, 179, 181, 185, 198
De Cruz, Helen 2–3, 7, 39–61
de Finetti, Bruno 225, 229, 233–5, 244
De Finetti table 233–5
Douven, Igor 2, 8–9, 173–99, 231, 233
Du Bois, W. E. B. 32
Dutilh Novaes, Catarina 2, 7, 17, 63–93

Edgington, Dorothy 234
empirical philosophy 1–2, 6–9, 15–27, 30–1, 65, 69, 96, 117, 119, 146–8, 163, 169
Equation, the 228–34, 237, 239, 241
experimenter's regress 25, 32
expertise 3–4, 7, 17, 102, 119
explication 31, 71

Familienähnlichkeit 31, 71
Fluge, Frithjof 168

formal correctness condition 148
Fraser, Alexander 169
Frege, Gottlob 145
fuzzy logic 175
fuzzy mathematics 175
fuzzy set theory 8–9, 173–98

Gärdenfors, Peter 9, 176, 180–2, 197, 255, 273
Gauss's Theorem 115
Gendler, Tamar 164
generalization problem 169
graded membership 174–9, 184, 187–8, 190, 195, 198
graphical inference 119, 123–44
Greiffenhagen, Christian 29
Griffiths, Phillip 123
Gullvåg, Ingemund 1
Gupta, Anil 169

Hersh, Reuben 3–4, 86
Horwich, Paul 145, 165–9

indicators, deductive 204, 206, 211–13
indicators, hedging 204, 206, 211–12, 216–19
indicators, inductive 204, 206, 211–12, 214–19, 221
indicator words 9, 203–7, 212, 214–15, 217–18
indispensability argument 51
induction, mathematical 96, 98–103, 111–13, 115–19
inductive generalization 44, 111, 117–18
Inglis, Matthew 1–13, 28–30, 66–8, 82–5, 124, 129, 138
internet study 129, 169
invariantism 39, 45, 50–3, 56

Jamnik, Mateja 100, 112
Jeffrey, Richard C. 235
Jeffrey table 235

Kamp, Hans 176–8, 183–4, 187–8, 191, 193, 195, 197
Kirkham, Richard 168
Knobe, Joshua 17–18, 28, 221
Kulpa, Zenon 98, 118, 124–5

Larvor, Brendan 8, 124–7, 136, 138–9
Lewis, David 9, 227–9, 239, 242, 251–4, 258–64, 266–7, 269–70, 272–3
Livengood, Jonathan 18, 146
Löwe, Benedikt 2, 3, 6–7, 15–37

Machery, Edouard 155, 274
Makinson, David 255
material adequacy 145–6, 149, 168
mathematical practice 2–3, 6–7, 15–18, 21, 27, 63, 65–6, 73–4, 79, 85, 88, 97–8, 123–6, 128–9, 135–6, 138–9, 213–14
mathematical practice, variations in 138
mathematical realism 7, 39–44, 51–6, 72, 86
mathematics education 1–3, 16–17, 29, 127–8, 137, 139
Mejía-Ramos, Juan Pablo 8, 28–9, 119, 123–44
Menger, Carl 22
mental model theory 9, 225–7
metaphilosophy 201, 219–20
Methodenstreit 22
methodology 17–18, 20–2, 26–7, 82, 85, 129, 168, 202, 206, 219–21
metrical inference 8, 124–9, 131–9, 142–4
Millean empiricism 39, 54–6
Mind 169, 207
Mississippi, University of 169
mixed method research 26, 252
Mizrahi, Moti 9, 201–24
modus ponens 164, 239
modus tollens 164
moral realism 7, 39–43, 52, 56
M-Turk (Amazon's Mechanical Turk) 156–7, 160, 169, 259, 264–5, 270, 274
Müller-Hill, Eva 17, 27–8
multidimensional scaling 180, 193

Næss, Arne 1, 8, 145–8, 150–5, 157–60, 163–4, 167–9
necessity 106, 109, 111, 116, 230
new paradigm 239, 241, 243–4
Nichols, Shaun 28

non-metrical inference 8, 124, 126–9, 131–9, 142–4
Norwegian undergraduates 150
numerical cognition 3–4, 7, 39, 44–7, 49–55
Núñez, Rafael 8, 27, 95–121

object file system (OFS) 43–4, 46–9, 51, 54–5
obvious 9, 201–21
operationalization 23–6, 28, 31, 176, 195, 260
Oslo, University of 150
Oslo Group 1, 6
Over, David 9, 225–49

Paolacci, Gabriele 169
Paris Congress of 1935 147
Partee, Barbara 176–8, 183–4, 187–8, 191, 193, 195, 197
philosophy of mathematical practice 2–3, 6–7, 15–18, 21, 27, 85, 135
Philosophy of Mathematical Practice, Association for the 16
platonism 4, 42, 52, 54, 86
probability, logic of 225, 238–9, 244
probability, subjective 225, 227, 231, 234, 238–9
prototype theory 176–8, 181, 199

Qualtrics 155–6, 169
Quine, Willard van Orman 6, 167, 202, 207, 251–2

Ramsey, Frank P. 145, 225, 228, 234, 244, 254
Ramsey test 9, 228, 234–5, 238, 251, 254–6, 272–3
real analysis 8, 124, 125–39
reasoning, conditional 225–44
reasoning, inductive 97, 104–6
reasoning, psychology of 1–2, 5, 9, 164, 225, 236, 238–9, 244
reasoning, visual 8, 96–8, 103, 117–18, 124–5, 136
Relaford-Doyle, Josephine 8, 95–121
replication 8, 18, 26, 159, 163
Ripley, David 4
Rynin, David 168

set theory 8–9, 43, 52, 56, 173–7, 179, 187, 195, 197
Shaffer, Michael J. 9, 251–76
shape space 179–80, 192–3, 195–7
Stalnaker, Robert 9, 227–9, 233, 239, 241–2, 251–4, 258–61, 264–7, 269–73
Street, Sharon 40–1
subitizing 43–4
synonymy/synonymity 150–5, 158–9, 163–4, 167
Sytsma, Justin 18, 146

Tarski, Alfred 145–50, 152–3, 155–8, 160, 163–7, 207
text mining 2, 201
Thomson, Daniel Greenleaf 169
triangulation, methodological 6, 15, 26–9, 32
truth 1, 6, 8, 27, 41, 52–4, 56, 75, 80, 98–9, 102, 112–13, 117, 145–58, 163–9, 180, 198, 202, 207, 215–17, 225–6, 228–30, 233–7, 241, 243–4, 253–4, 263–4, 268–9, 271–3
truth, Aristotelian conception of 146
truth, concept of 145–50, 152–3, 155, 157–8, 165–7, 169
truth, deflationary account of 8, 157, 165–6, 168
truth, Horwich's minimalism about 145, 165–9
truth, nonphilosopher's notion of 146–9, 152, 155, 158, 163, 165–7
truth, Tarski's semantic conception of 145–50, 152–3, 155–8, 160, 163–7

Ulatowski, Joseph 1, 8, 145–71
uncertainty 236–7, 240
universal quantifier 106, 110, 119

vagueness 9, 175–7, 197–8, 252, 257–8
validity, probabilistic 236, 238, 241
van Fraassen, Bas 252
Van Kerkhove, Bart 2, 6–7, 15–37
visual proof 8, 77, 95–106, 109, 111, 113, 115–19, 124–5, 136
visual proof by induction 8, 95–6, 98–103, 111, 113, 115–19
von Schmoller, Gustav 22
Voronoi tessellation 180–7

Waikato, University of 169
wave theory of light 153, 158
Weber, Keith 3, 8, 28–9, 119, 123–44
Wittgenstein, Ludwig 31, 86

Yale Experiment Month 155

Zadeh, Lotfi 174–5, 195–6
Zhen, Bo 119, 127, 129, 137, 139

www.ingramcontent.com/pod-product-compliance
Lightning Source LLC
Chambersburg PA
CBHW070020010526
44117CB00011B/1648